SpringerBriefs in Molecular Science

History of Chemistry

Series Editor

Seth C. Rasmussen, Fargo, USA

For further volumes:
http://www.springer.com/series/10127

Jay A. Labinger

Up from Generality

How Inorganic Chemistry Finally
Became a Respectable Field

 Springer

Jay A. Labinger
Beckman Institute
California Institute of Technology
Pasadena
USA

ISSN 2212-991X
ISBN 978-3-642-40119-0   ISBN 978-3-642-40120-6   (eBook)
DOI 10.1007/978-3-642-40120-6
Springer Heidelberg New York Dordrecht London

Library of Congress Control Number: 2013945786

Printed on acid-free paper

Springer is part of Springer Science+Business Media (www.springer.com)

Acknowledgments

I am grateful to Seth Rasmussen, series editor, for supporting this work and helping in all sorts of ways. I thank a number of people for help with finding, obtaining, and/or analyzing documents, photos and other materials: Andrew Mangravite and David Caruso at the Chemical Heritage Foundation; Jeff Carroll with the Gordon Research Conferences; Farai Tsokoday and Sara Rouhi at the American Chemical Society; Martin Jansen of the *Zeitschrift für Anorganische und Allgemeine Chemie*; Margaret Janz at Indiana University; Bill Griffiths at Imperial College London; Chris Blackman at University College London; Bill Connick and Bill Jensen at University of Cincinnati; Rich Jordan at University of Chicago; Alison Butler and Ralph Pearson at UC Santa Barbara; John Shapley and Vera Mainz at UIUC; Kathy Armbruster at University Science Books; Diana Kormos Buchwald, historian of science at Caltech; and particularly Shelley Erwin and Loma Karklins in the Caltech archives. Special thanks go to Dana Roth, Caltech chemistry librarian extraordinaire, for pointing me in the right direction innumerable times.

I am greatly indebted to those who were kind enough to participate in interviews about their personal recollections: Ted Brown, Mel Churchill, Jim Collman, Rich Eisenberg, Jack Halpern, Dick Holm; and several Caltech chemists: Fred Anson, Jack Roberts, and above all Harry Gray, without whose ongoing encouragement (and inexhaustible store of inside information) this project would have been much less enjoyable.

Contents

Abstract

Inorganic chemistry, with a negation in its very name, was long regarded as that which was left behind when organic and physical chemistry emerged as specialist fields in the nineteenth century. Scarcely differentiated from general chemistry, inorganic chemistry was not widely accepted as an independent, intellectually viable discipline, especially in US academia, before the middle of the twentieth century; only then did it begin to gain its current stature of equality with that of the other main branches of chemistry. Discussion of the evidence for this transition, both quantitative and anecdotal, includes consideration of the roles of local and personal factors, with particular focus on the Chemistry Division at the California Institute of Technology, as an illustrative example. Examination of key developments, as well as the central figures that fostered them, leads to proposed explanations for the remarkable upgrade of status enjoyed by inorganic chemistry.

Keywords Inorganic chemistry · History of chemistry · Discipline formation · American inorganic chemists · Chemistry in US academia · Chemistry at Caltech · Mechanism in chemistry · Organometallic chemistry · Donald Yost

Chapter 1
Introduction

> *"Chaplain, I once studied Latin. I think it's only fair to warn you of that before I ask my next question. Doesn't the word Anabaptist simply mean that you're not a Baptist?.... Now, Chaplain, to say you're not a Baptist doesn't really tell us anything about what you are, does it? You could be anything or anyone."*
>
> Joseph Heller, Catch-22

In the late summer of 1967, between my junior and senior years at Harvey Mudd College (HMC) in California, I traveled east to look at some possible choices for graduate study in chemistry. At Harvard I stopped by the chemistry department office and asked if any faculty members were available to talk about the graduate program. The head of the office staff (she had been in that position for many years, and was clearly used to such inquiries) replied, graciously, "Certainly. Are you organic or physical?"

Her question, thus phrased, surprised me considerably, as I was leaning towards specializing in inorganic chemistry for graduate work. I had done a summer project, and was planning to do my senior research project, in that subfield (with Mits Kubota, the inorganic chemist at HMC). It had never occurred to me that its status might be considered inferior elsewhere. When I told her my preference, she seemed at least equally surprised, but quickly recovered, and arranged a visit that, while impressive, was mostly limited to organic chemists.

I chose to go to Harvard anyway, and to do my Ph.D. in inorganic with John Osborn (Fig. 1.1), who arrived at Harvard shortly after my visit. That went very well; but I had a few more disconcerting experiences during my first year. The inorganic faculty, besides Osborn, consisted of full professor Eugene Rochow, who was slated to retire the following year, and Mel Churchill, another assistant professor. A search for a full professor to succeed Rochow brought in several of the leading names in the field—Harry Gray, Fred Basolo, Fred Hawthorne—each of whom spent a week at Harvard and gave what seemed to me several first-rate lectures. In the end, though, none was offered the position, which was left vacant for a number of years thereafter. I found opportunities to ask a couple of the senior organic faculty why, and was told (they were quite open about their opinions!) the department felt that not only the particular candidates interviewed, but the field of inorganic chemistry as a whole, fell well short of the standards of intellectual importance and respectability that Harvard required of its senior appointees.

As I pursued my career in inorganic chemistry over the next few decades, it was eminently clear (to me, anyway) that the field *did* enjoy a status fully coequal to

J. A. Labinger, *Up from Generality*, SpringerBriefs in History of Chemistry,
DOI: 10.1007/978-3-642-40120-6_1, © The Author(s) 2013

Fig. 1.1 John Osborn (1939–2000) with his students at Harvard in 1971. From *left*: Al Kramer, Sue Bezman, Steve Wilson, the author, John Shapley, Dave Rice (*kneeling*), Dick Schrock. Osborn left Harvard (putting him in good company: see Chap. 4) for Strasbourg in 1975, where he continued his outstanding work in organotransition metal chemistry until his untimely death (Image courtesy of John Shapley)

that of any of the other subdivisions of chemistry, and I wondered about the origins of Harvard's disdain. Was it just a local, idiosyncratic consequence of the undeniably immense strength of their tradition and faculty in organic and physical chemistry? Or did it have deeper and broader roots? Had inorganic chemistry previously been *generally* regarded as a lesser field? If so, when and how did this attitude change (except at Harvard)?

Then, too, the question of what "inorganic chemistry" really means came into play. The very name is somewhat disturbing: a discipline defined by negation, by what it is *not*. As one practitioner of the field comments:

> Inorganic chemistry is a subject that exists by default—it is the part of chemistry that remained when organic chemistry (the chemistry of carbon compounds containing at least some carbon-hydrogen bonds) and physical chemistry (the science of physical measurements as applied to chemical systems) developed as distinct subdisciplines in the nineteenth century [1].

That seems quite a negative characterization: it's as though inorganic chemistry is not really a subdiscipline at all, but just the dregs that were left over after all the interesting topics were demarcated and separated out. Historically, that is not entirely inaccurate. When first organic and then physical chemistry developed as specialist disciplines, in the second half of the nineteenth century, inorganic chemistry was left to become more or less conflated with general chemistry. Consequently—much as general practitioners in medicine tend to be underappreciated relative to specialists—for a long time inorganic chemists lagged far behind their more glamorous organic and physical colleagues in generating interest in and respect for their efforts.

Unquestionably, however, that is no longer the case: inorganic chemistry eventually came to take its place as a specialist subdiscipline, distinct from general chemistry, and competitive by any standard of metrics with the longer-established fields of organic and physical chemistry. On one website we can even find the statement "Inorganic chemistry is the most important area of chemistry" [2]![1] At some point, therefore, the perception of the field must have undergone a major alteration. The central theme of this book is to locate that transition in time—I place it around the middle of the twentieth century—and to trace the key developments and people that were primarily responsible for it.

In the balance of this introductory chapter I briefly summarize the separation of organic and physical chemistry as specialist subdivisions, and the resulting state of inorganic chemistry. Chapter 2 discusses developments in the latter field through the early twentieth century, which included some important advances but did not accomplish any significant re-evaluation of its status. Chapter 3 presents evidence—both statistical and anecdotal—that a major change did in fact take place during the 1950s, at least within the USA. In Chap. 4 I digress from examining general trends to explore how local factors and personalities influenced developments at one particular institution (my own at present, Caltech). Chapter 5 returns to the global view, identifying the most important topics and researchers that made inorganic chemistry respectable; and Chap. 6 closes by trying to account for how inorganic chemistry has been able to sustain a coherent, intellectually viable persona for itself in the face of increasing, seemingly centrifugal interdisciplinary forces.

The history of the very term "inorganic chemistry" is, not surprisingly, somewhat indistinct. "Inorganic" and "organic" were applied to substances, to distinguish those of mineral[2] versus animal and vegetable origin, at least as far back as the eighteenth century; one historian attributes the first distinction between "non-organic and organic bodies" to Bergmann in 1775 [3]. But such considerations do not appear to have been much used by chemists (although Bergmann *was* a chemist!) until the nineteenth century [4]. Russell describes textbooks written by Berzelius in the early nineteenth century as divided into sections on inorganic, vegetable and animal chemistry, with the term "organic chemistry" emerging around 1810 [3]. The earliest reference in the Oxford English Dictionary in which the word "inorganic" is associated with the word "chemistry" dates from 1831, while the earliest that includes the actual phrase "inorganic chemistry" is 1837 [5]; another reference also places the first use of that phrase in the 1830s [6].

It is still less clear when the term "inorganic chemist" began to be used. Certainly, at the beginning of the nineteenth century there was no separation of chemistry into subfields; there were just chemists. But from the middle of the century onwards, most of the crucial developments in understanding and systematizing chemistry were coming from studies of organic substances, and organic chemistry emerged—not just as a recognizably distinct specialization, but

---

[1] It should be noted, however, that this quote comes from an essay prefaced by: "**Warning!** The following article is from The Great Soviet Encyclopedia (1979). It might be outdated or ideologically biased."

[2] *Chimie minérale* continues to be commonly used in French.

as the "clearly dominant" and even "hegemonic" area of chemical research [7]. Typical characterizations of this period by historians of chemistry, ranging from the early twentieth century[3] to much more recent work, include the following:

> Between 1870 and 1890 the rapid development of organic chemistry gave it such a relative prominence that the other branches of the science rather suffered in consequence. Inorganic chemistry particularly seemed to be drifting towards the discouraging position of a completed science, and some predicted for it little further growth....interest centered upon the refinement of analytical procedures, the discovery of new elements, and the revision of atomic weight. Such researches were of course of great value but, from the historic point of view, they leave little to record, because they introduce little which is new in the way of important principles [8].

> While the great development of organic chemistry was taking place, a smaller number of chemists continued to devote themselves to the older discipline of inorganic chemistry. Some of these men also worked in the new field of physical chemistry.... As a result of all these factors the foundation for great progress in *general* chemistry were laid down during the nineteenth century [9]. (My italics; we will come to "the new field of physical chemistry" shortly.)

> At the beginning of the nineteenth century, the unity of the discipline had been assumed because the laws of inorganic chemistry extended to compounds produced by organic beings. In the 1850s the situation reversed itself. It was organic chemistry that provided concepts with which to study and classify the inorganic realm [10].

> Organic chemistry was riding high.... Inorganic chemistry, which under Berzelius had been the dominant branch, was beginning to look like a poor relation [11].

In such an environment, clearly, associating with the dominant field by defining oneself as an organic chemist was a prestigious move, whereas there was no such advantage in being known as an inorganic chemist. Also, it appears that the low nineteenth century esteem for the field has continued to color more modern historical viewpoints: historians of chemistry have indeed tended to find relatively "little to record" of inorganic researches. Of the 20 or so more-or-less comprehensive histories of chemistry that I have looked at, only a few include a separate chapter or section on inorganic chemistry—often the phrase even fails to appear in the index—while the word "inorganic" never appears *at all* in two essays on the history of chemistry [12, 13]! A collection of important papers from the first half of the century [14] divides them among four topic headings—techniques, general and physical chemistry, organic chemistry, and biochemistry—inorganic is conspicuously absent. Likewise, a set of essays on twentieth century chemistry [15] includes chapters on organic, theoretical, radio- and nuclear chemistry, geo- and cosmochemistry, biotechnology, polymer chemistry, and materials; but the word "inorganic" appears nowhere in the table of contents (or, again, in the index). (It is

---

[3] The publication data of the edition of this work that I consulted is actually 1931, but it is a revision (not by the original author) of an earlier (1918) edition; the section dealing with inorganic chemistry does not appear to have been revised much if at all, as there are no references to developments later than 1918.

true, of course, that inorganic chemistry is intimately connected with many of the topics that *are* covered in both of these collections, but the disinclination to name it is revealing.) A survey of chemical history studies appearing from 1985 to 2005 [6] devotes a measly eight pages (out of well over 200) to inorganic chemistry. The index to 67 years worth of issues of the journal *Ambix* refers to 32 articles on organic chemistry, 18 on physical chemistry, zero on inorganic chemistry [16].

What made inorganic chemistry such a "poor relation?" First, as the above-quoted comments suggest, developments in organic chemistry were increasingly explanatory, while inorganic chemistry remained predominantly descriptive and phenomenological. According to Nye "organic chemistry developed a program of study, a language of discourse, and a system of explanation that was foreign to the practitioners of an earlier *general* chemistry" (again, my italics) [17]. Similarly, Russell comments that in the 1860s, concepts of structure arising from organic studies emerged as a key organizing principle, while the main problem for inorganic chemists was uncertainties in atomic weight determinations [18]—an important issue, no doubt, and vital to subsequent progress, but not exhibiting much of an obvious intellectual or epistemological component. Another contributing factor, I propose, was the inability of inorganic chemistry during this period to distinguish itself from general chemistry (note the references to the latter term in two of the preceding quotations), so that chemists felt little need to associate themselves with a *non*-specialty, while historians (and others) felt little need to explicitly label important developments with the term "inorganic." (I say "during this period," but as we shall see in the next chapter, that conflation of inorganic and general chemistry remains substantially unchanged, well into the twentieth century.)

As one illustration of the failure of inorganic chemistry to gain much traction, consider the mid-nineteenth century work of Edward Frankland (Fig. 1.2). His studies of organometallic compounds, particularly the zinc alkyls (ZnR_2), played a major role in developing the concept of valence. One might expect that such a discovery could and should have served as "a concrete argument in favor of a reunion of organic and inorganic chemistry at the moment when the divorce was the most bitter" [19]. But no such reunion took place, and Frankland's findings remained primarily in the organic realm. Organometallic chemistry *did* eventually have a major impact in inorganic chemistry—but not until a century later, as we shall see in Chap. 5.

Towards the end of the nineteenth century the preeminence of organic chemistry *was* finally challenged—but not by inorganic chemists; instead physical chemistry emerged as a new specialized subfield. That development has been extensively discussed, both as an important stage in the history of chemistry and as a revealing case study in disciplinary formation. We will return to the latter aspect in Chap. 6; for now it is worth noting that some of the central figures thought of the establishment of their new field as a *unifying* force, with the potential to re-integrate organic and inorganic chemistry [7, 17]. Ostwald, for example, originally sought to call it *allgemeine Chemie* "for in his view it would not be a new part of chemistry so much as a new basis for all existing parts of the science" [20].

Fig. 1.2 Edward Frankland (1825–1899), looking rather youthful (date unknown). Frankland was a leading English organic chemist of the nineteenth century; he pursued his career first in Germany and then in England, eventually at the Royal Institution (Image downloaded from Wikimedia Commons, in the public domain)

Another member of the Bunsengesellschaft (a professional association of physical chemists) called in 1899 for reversing the deplorable "abandonment of inorganic chemistry" at the expense of organic chemistry [21].

That didn't happen. "Contrary to what one might have expected, perhaps, the traditional structures now became hardened and the newcomer, instead of healing the breach, actually helped to extend it" [22]. As a mid-nineteenth century author wrote: "The most important feature in the recent progress of Inorganic Chemistry has been the rigorous verification which numerical data of all kinds have received, whether relating to physical laws, such as the specific heat of substances, or to chemical properties and composition" [23]. It was, of course, precisely those aspects that formed the nucleus of the new field. The creation of physical chemistry effectively served to carve away the most scientifically respectable pieces of what organic chemistry had left behind, thereby reducing inorganic chemistry at the end of the nineteenth century to a still less impressive residue.

References

1. Swaddle T (1997) Inorganic chemistry: an industrial and environmental perspective. Academic Press, San Diego, p 1
2. http://encyclopedia2.thefreedictionary.com/inorganic+chemistry. Accessed 11 March 2013
3. Russell CA (1976) The structure of chemistry, Unit 1. Open University Press, Milton Keynes
4. Levere TH (2001) Transforming matter: a history of chemistry from alchemy to the Buckyball. Johns Hopkins Press, Baltimore, p 95
5. Oxford English Dictionary (1933/1961) Oxford University Press, London

6. Campbell WA (2005) Inorganic Chemistry. In: Russell CA, Roberts GK (eds) Chemical history: reviews of the recent literature. RSC Publishing, Cambridge, UK, p 49

7. Nye MJ (1993) From chemical philosophy to theoretical chemistry: dynamics of matter and dynamics of disciplines, 1800-1950. University of California Press, Berkeley, CA

8. Moore FJ (1918/1931) A history of chemistry (revised by W. T. Hall). McGraw Hill, New York, p 237

9. Leicester HM (1956) The historical background of chemistry. John Wiley & Sons, New York, p 189

10. Bensaude-Vincent B, Stengers I (1996) A history of chemistry (translated by D. van Dam). Harvard University Press, Cambridge, MA, p 127

11. Levere TH (2001) Transforming matter: a history of chemistry from alchemy to the Buckyball. Johns Hopkins Press, Baltimore, p 148

12. Thackray A, Sturchio JL (1987) What's past is prologue: two hundred years of chemistry in America. J Electrochem Soc 134: 558C–564C

13. Servos JW (1985) History of chemistry. Osiris 1:132–146

14. Leicester HM (ed) (1968) Source book in chemistry, 1900–1950. Harvard University Press, Cambridge, MA

15. Reinhardt C (ed) (2001) Chemical sciences in the twentieth century: bridging boundaries. Wiley-VCH, Weinheim

16. Ambix: contents and index, 1937–2003 (2004)

17. Nye MJ (1996) Before big science: the pursuit of modern chemistry and physics, 1800–1940. Twayne Publishers, New York

18. Russell CA (1976) The structure of chemistry, Unit 2. Open University Press, Milton Keynes

19. Bensaude-Vincent B, Stengers I (1996) A history of chemistry (translated by D. van Dam) Harvard University Press, Cambridge, MA, p 149

20. Servos JW (1990) Physical chemistry from Ostwald to Pauling: the making of a new science in America. Princeton University Press, Princeton, p 4

21. Barkan DK (1999) Walther Nernst and the transition to modern physical science. Cambridge University Press, Cambridge, UK, p 16

22. Russell CA (1976) The structure of chemistry, Unit 3. Open University Press, Milton Keynes

23. Graham T (1850/1858) Elements of inorganic chemistry (2nd revised edition, ed H. Watts and R. Brides). Blanchard & Lea, Philadelphia, p vii

Chapter 2
False Labor: Inorganic Chemistry in the Late Nineteenth-Early Twentieth Centuries

> *Those of us who were familiar with the state of inorganic chemistry in universities twenty to thirty years ago will recall that at that time it was widely regarded as a dull and uninteresting part of the undergraduate course….that the opportunities for research in inorganic chemistry were few, and that in any case the problems were dull and uninspiring; as a result, relatively few people specialized in this subject.*
> Ronald Nyholm, *The Renaissance of Inorganic Chemistry* (1956)

The previous chapter paints a rather bleak portrait of inorganic chemistry in the late 1800s. However, both chemists and historians have pointed to a significant upgrade in status—using terms such as revival, rebirth, renaissance—taking place even before the turn of the century. One historian commented in 1906: "If we glance back over the labors of the last 50 or 60 years, we recognize that organic chemistry has gone on preponderating more and more over inorganic…. A review of the chemical literature of the last 10 or 20 years shows very clearly the revived influence of inorganic chemistry as an incentive to research" [1]. H. N. Stokes, an important American chemist of the time,[1] published a lengthy *Science* article in 1899 entitled "The Revival of Inorganic Chemistry" [2]. (It should be noted, however, that he followed it with "The Revival of Organic Chemistry" [3], acknowledging that many might find *that* topic "almost facetious.") There is no question that many important advances that took place in the late nineteenth through the first part of the twentieth centuries played a crucial role in the evolution of inorganic chemistry as an independent subfield. Nonetheless, it would be incorrect to apply so momentous a term as "rebirth" to this early period.

Before surveying those developments, I offer a brief comment on the appropriateness of terms. It is true that research in what we would now call inorganic chemistry languished during the greater part of the nineteenth century and began to grow again towards the end. However, it does not seem quite accurate to call that a "*re*birth" or "renaissance" of the *field* of inorganic chemistry. Such a field never existed as a distinct entity; the earlier researches were part of chemistry *tout court*.

---

[1] Henry Newlin Stokes (1859–1942) was a chemist with the US Geological Survey and Bureau of Standards, and served a term as President of the ACS around the turn of the century, before turning to philosophy, becoming a leader of the Theosophical Society.

J. A. Labinger, *Up from Generality*, SpringerBriefs in History of Chemistry, DOI: 10.1007/978-3-642-40120-6_2, © The Author(s) 2013

Fig. 2.1 Dmitry Mendeleev (1834–1907), date unknown. His discovery of the Periodic Law is arguably the most important (eligible) contribution in chemistry that was never recognized with a Nobel Prize (Image courtesy of the University of Pennsylvania Library's Edgar Fahs Smith Memorial Collection)

Fred Basolo makes a similar point, with regard to similar mid-twentieth century characterizations: "Everyone talks about the renaissance of inorganic chemistry…. Actually, I'm inclined to call it the "birth" of inorganic chemistry because renaissance means that you're coming back to something that has already been done" [4]. Be that as it may, I will bow to common practice and continue to apply "revival", "renaissance," etc. to all (real and/or perceived) upgrades in status.

The first (and foremost) advance, of course, was the work culminating in the 1870s with Mendeleev's (Fig. 2.1) Periodic Table. Many historians have proclaimed its significance: "It is with the construction of the Periodic Table that the story of 1800s inorganic chemistry begins" [5]. "[Inorganic chemistry], so long over-shadowed by organic chemistry, so long but little more than a collection of almost un-connected facts, subordinate to analytical and technical chemistry and to mineralogy, is gradually, and especially since the discovery of the Periodic Law, rising to the rank of an independent and important division of our science" [3]. "The Periodic Law…stimulated the study of Inorganic Chemistry, which had been rather neglected in the second half of the nineteenth century owing to the great specialization in Organic Chemistry" [6].

All of that is true; but in the last source cited we also read "The development in *general* chemistry during the twentieth century originated in the Periodic Law" (my italics) [6]. The Periodic Table is unquestionably a *sine qua non* for any systematic study of the chemistry of the elements; but its evolution really belongs to general, not inorganic, chemistry. Or, better put, it solidified the *conflation* of general and inorganic chemistry, rather than advancing inorganic chemistry as a distinct, respected subfield. Mendeleev himself held the title of "Professor of

General (Inorganic) Chemistry" at the University of St. Petersburg [7]! Russell comments: "It [the Periodic Table] also provided for inorganic chemistry its first great generalization....But it is all too easy to overstate its importance for suggesting lines of research....Indeed, it is not going too far to say that the most important discoveries in inorganic chemistry for the rest of the century not only owed little to the Periodic Table but actually offered it an embarrassing challenge" [8].

In any case, the preponderance of late nineteenth century inorganic studies, though more systematic than before Mendeleev, remained largely descriptive and phenomenological. There was not much interest or activity in the more explanatory mode that organic chemists had established. "In 1910 many specialists in inorganic chemistry still thought that the atomic and molecular hypothesis was only a fiction....Although molecular structures had had an impact on organic chemistry, they had remained relatively peripheral in inorganic chemistry, which was more concerned with the variety of elements that entered into compounds than with the structures built by molecules" [9].

To be fair, it must be acknowledged that many of the founders of physical chemistry were at least equally skeptical of the utility, let alone the reality, of atoms and molecules. But those skeptics *did* construct their science upon an alternative philosophical framework, based on energetics [10], whereas inorganic chemistry of the time had little in the way of comparable intellectual underpinnings to offer. The one exception—and the most significant advance in the field around the turn of the twentieth century—is to be found in the work of Alfred Werner (Fig. 2.2).

Werner has been claimed as a national by the Germans, French and Swiss. He was born in Mulhouse while Alsace was still part of France, remained there after it was seized in the Franco-Prussian War of 1870 (and even served in the German army); but he spent most of his career as an independent researcher in Zurich, coming to the University of Zurich (initially as an organic chemist!) in 1893. By then he had already taken an interest in coordination compounds, which at the time were poorly (if at all) understood, especially with regard to constitution and structure [11]. The existence of a number of series of species, each containing a metal in combination with the same constituents but in varying numbers, was very hard to reconcile with well-established laws of proportions and valency.

For example, cobalt-ammine-chloride compounds of formulae $Co(NH_3)_xCl_3$ were known for x = 3, 4, 5 and 6, all of different colors. Cobalt was considered to be trivalent, which was taken to mean that it could only bond to three entities; how could that be made compatible with the known compositions? Before Werner, the dominant model was that of Jørgensen,[2] who had proposed the chain structures shown in Fig. 2.3. These did correctly capture *some* of the known chemical

---

[2] Sophus Mads Jørgensen (1837–1914), a Danish chemist who made many of the early important *experimental* discoveries in coordination chemistry, but fought a long rear-guard action against Werner's conceptual interpretation, until finally acknowledging the latter's triumph in the early twentieth century.

Fig. 2.2 Alfred Werner
(1866–1919), the father of
coordination chemistry, at the
time he received the 1913
Nobel Prize in Chemistry
(Image downloaded from
Wikimedia Commons, in the
public domain)

behavior, notably the varying number of ionizable chloride ions. Those at the end
of the chains were assumed to be ionizable, whereas those attached directly to
cobalt were not. However, there were clear anomalies; the case where $x = 3$ (i.e.,
$Co(NH_3)_3Cl_3$) should behave much like that for $x = 4$ in Jørgensen's model, but
the former is entirely *non*-ionic, while the latter readily liberates one Cl^- ion [12].

In a groundbreaking series of papers beginning in 1893 [13], Werner completely reformulated these and related species. First he introduced a distinction
between groups directly bonded to the central metal atom and those affiliated only
by ionic forces. In due course these came to be called inner- and outer-sphere
interactions, respectively, with the former eventually termed "ligands" (by Stock
in 1916 [14]). He further recognized that the "magic" number was not three, the
valency (in traditional usage) of cobalt, but rather *six*, the characteristic "coordination number" of trivalent cobalt. The complexes of Fig. 2.3 would thus be
represented instead as $[Co(NH_3)_3Cl_3]$, $[Co(NH_3)_4Cl_2]Cl$, $[Co(NH_3)_3Cl]Cl_2$, and
$[Co(NH_3)_6]Cl_3$, where the Cl's outside the brackets are ionizable. He then
extended this concept to include spatial representation, postulating an octahedral
arrangement of six groups around a central atom as the obvious analog of the
organic chemist's tetrahedron, and observed that certain compositions should exist
as more than one structural isomer, as shown (for $x = 4$) in Fig. 2.4. Even more
dramatically, he recognized that certain arrangements of ligands could give rise to
the possibility of optical isomerism, a phenomenon previously deemed unique to
the organic realm. All of these predictions were already known (or were soon
shown) to be consistent with experimental finding [8, 12].

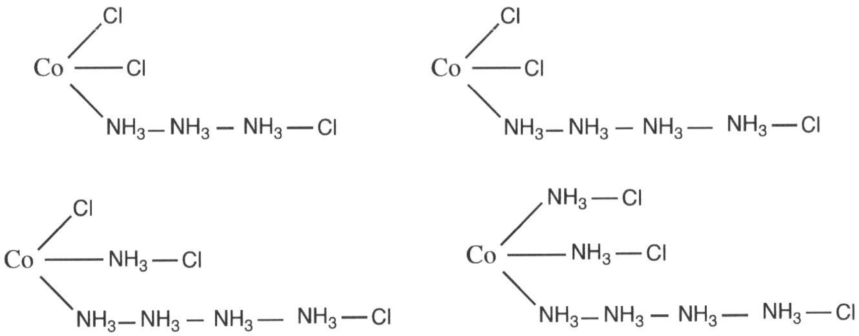

Fig. 2.3 Jørgensen's chain structure model for the $Co(NH_3)_xCl_3$ series

Fig. 2.4 The two isomers of [$Co(NH_3)_4Cl_2$]Cl (the ionic Cl's are not shown)

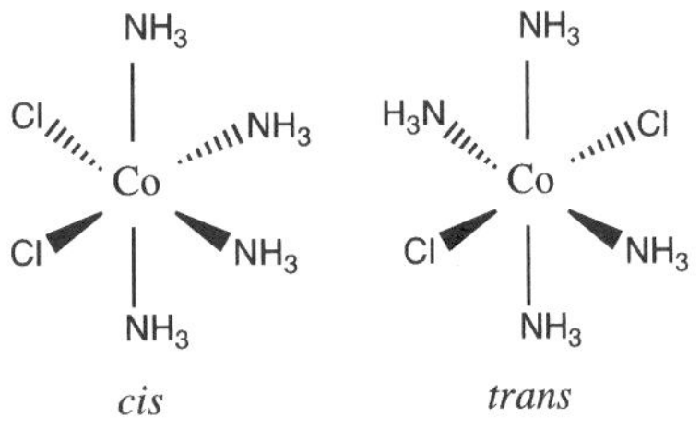

Werner's work *has* been characterized by some as an early renaissance (or, as Basolo would prefer, a "naissance") of inorganic chemistry. For example, "The progress in carbon chemistry outshone that made in other areas until the 1890s, when new discoveries and theories relating to coordination compounds signalled the coming of age of inorganic chemistry" [15]. On the other hand, Russell describes the period quite differently: "It was simply not true that coordination complexes played a key role in inorganic chemistry either then [just before the First World War] or for 40 years ahead. What Werner did do in his own fairly short lifetime was to convince people that *in this area*...his theory was a satisfactory explanation" (italics in the original) [8].

George Kauffman, a leading Werner scholar, has suggested that Werner's management style may have been part of the reason that his work did not ignite a major expansion of the field: "[O]ne might also wish to ponder whether it was [his] high degree of regulation and supervision which may have prevented the formation of a Werner school....Perhaps the impact of Werner's powerful, authoritarian personality and the impression of his control and mastery of his field deterred most of those who had worked with him from any thought of following in his footsteps" [16]. In any case, while there is no doubt that Werner's work lies at the very center of the emergence of inorganic chemistry as an intellectually respected field, the preponderance of evidence, as we shall see, shows that no such emergence took place until *much* later—about 40 years later, as Russell says.

Nonetheless, Werner's contemporaries were already beginning to show *some* renewed interest in the field. The most notable development was the founding of the first journal devoted to the topic, the *Zeitschrift für anorganische Chemie* (*ZAC*), in 1892. The very first issue includes an editors' note, offering a rationale for a new journal, which begins (my translation[3]) [17]:

> At present, reports of inorganic chemistry research submitted for publication are dispersed among a very large number of domestic and foreign journals; they appear as strangers among the ever-increasing number of works in the field of chemistry of carbon compounds. This situation is inconsistent with the current importance of inorganic chemistry, which in recent decades has emerged from the confines of its narrowly defined science to take part in the resolution of questions which are of great significance for chemistry in general.

It is not at all clear to me what they might have meant by "emerged from the confines of its narrowly defined science." In what way was the *purview* of inorganic chemistry constrained in the earlier nineteenth century, or less so towards the end? Examination of the technical content of that first issue doesn't help much; of the 34 articles a large fraction deal with matters that could well be considered at least as appropriate for "chemistry in general" rather than specifically inorganic. The two longest, by Harvard chemist T. W. Richards, are on determining the atomic weight of copper with greater precision; another is on the phenomenon of coal dust explosions! Like the Periodic Table, the introduction of this journal perhaps does more to blur any distinction between general and inorganic chemistry than to help secure the latter's standing.

To bolster that argument, we need only look at the subsequent history of the *title* of this journal. In 1915, it became known as *Zeitschrift für anorganische und allgemeine Chemie* (*ZAAC*: journal of inorganic *and* general chemistry); in 1943 it reverted to the original *ZAC*; and then in 1950, back again to *ZAAC*, which remains its title to this day. According to one of the current editors, the first shift was largely the initiative of the editor at that time, who was personally very interested in a broad range of topics such as metallurgy, and induced the publishers to change the name; the second was brought about by the president of the Deutsche Chemische Gesellschaft, who wanted to keep inorganic and physical chemistry strictly separate, and was able to force his preferences upon the editors by virtue of his good connections with the Nazi regime. After the war, that move was reversed [18].

To be sure, *ZAC* (and later *ZAAC*) did publish many papers that play an important part in the continuing development of inorganic chemistry, including much of Werner's early work, as well as a good deal of the next major European figure in inorganic chemistry, Alfred Stock (1876–1946). But Werner ceased publishing in *ZAC* in 1899 (after that date most of his work appeared in

---

[3] "Die Mitteilungen über anorganisch-chemische Untersuchungen sind bis jetzt in einer sehr grossen Ansahl von in- und ausländischen Zeitschriften verstreut zur Veröffentlichung gelangt; sie erscheinen als Fremdlinge unter der immer mehr wachsenden Anzahl von Arbeiten aus dem Gebiete der Chemie der Kohlenstoffverbindinungen. Diese Stellung entspricht nicht der heutigen Bedeutung der anorganische Chemie, denn diese ist im Laufe der letzten Decennien aus dem engen Rahmen einer rein beschreibenden Naturwissenschaft herausgetreten und nimmt Teil an der Entscheidung von Fragen, welche für die allgemeine Chemie von hoher Bedeutung sind."

Berichte [11]), and resigned from its editorial board, feeling that the journal had moved away from inorganic chemistry in favor of physical chemistry [19]. (On the other hand, Stock published almost exclusively in *Berichte* until around 1925, only after then using *ZAC* for a portion of his output.)

Stock enjoyed a long career of investigations into the chemistry of main group elements, primarily boron and silicon, at several German universities (Breslau, Berlin, Karlsruhe), and developed much of the methodology needed to work with such species, many of which are volatile, air-sensitive and/or toxic. He was particularly noted for his introduction of vacuum line techniques (which, unfortunately for him, entailed usage of large quantities of mercury: he suffered terribly from mercury poisoning for the last several decades of his life). But even a (rather hagiographic) scientific biography does not credit his work as amounting to a renaissance; rather it proposes that "his own life's work…laid the sure foundation of a *future* renaissance in inorganic chemistry" (my italics) [20].

Of course there were other important inorganic chemists during the first half of the twentieth century, most of them also in Germany. I will not undertake an extensive survey, but will just mention Walter Hieber (1895–1976), whose research program, primarily at the Technische Hochschule München, essentially created the topic of metal carbonyl chemistry, bringing it from a small handful of "peculiar compounds" to a highly populated class of metal complexes, extending to virtually all the transition metals. Like Stock's work, Hieber's studies played a crucial role in subsequent developments [21].

Outside of Germany we find much less evidence of revitalized interest in the field. Of course there *were* advances during this period that, like the Periodic Table, were absolutely essential for further progress in systematizing inorganic chemistry, particularly the contributions to the understanding of the nature of chemical bonds made by two American chemists: Gilbert N. Lewis (1875–1946) and Linus Pauling (1901–1994) [22]. Their work was seminal: one chemist/ historian recalls that a popular exam question in the years before Pauling was "Is inorganic chemistry a largely closed and finished subject?" [23]. But neither of these giants was really associated with the field of inorganic chemistry. Lewis always called himself a physical chemist, while Pauling's title at Caltech, which varied over the years, never included any reference to inorganic chemistry. Indeed, he said that any interest he had in inorganic (and organic, for that matter) chemistry was "almost entirely from the structural point of view" [24].

Like the other milestones examined in this chapter—the Periodic Table, Werner's work on coordination compounds, the establishment of the first dedicated journal, Stock's work on main group compounds—the work of Lewis and Pauling did not result in any *immediate* improvement of the status of inorganic chemistry. To be sure, that *was* to come, as a farsighted turn-of-the-century commentator opined: "It is not to be expected, nor is it to be desired, that inorganic chemistry will at once sweep organic chemistry from its position of preeminence. The causes to which this is due may outlast our generation, but that the inorganic tide is rising, and that this branch will finally attain its due position, cannot be doubted" [2]. But any birth (or rebirth) announcement of inorganic chemistry before the middle of the twentieth century would have to be considered as decidedly premature.

References

1. von Meyer E (1906) A history of chemistry from earliest times to the present day (3rd English edition, translated by G. McGowan). Macmillan & Co., London, p 417
2. Stokes HN (1899) The revival of inorganic chemistry. Science 9: 601–615
3. Stokes HN (1900) The revival of organic chemistry. Science 12: 537–556
4. Basolo F (2002) Fred Basolo interview by Arnold Thackray and Arthur Daemmrich at Northwestern University, Evanston, Illinois, 27 September 2002 (Philadelphia: Chemical Heritage Foundation, Oral History Transcript # 0264)
5. Cobb C, Goldwhite H (1995) Creations of fire: chemistry's lively history from alchemy to the atomic age. Plenum Press, New York, p 257
6. Partington JR (1957/1989) A short history of chemistry (3rd ed). Dover, New York, chapter XV
7. Brock WH (1992) The Norton history of chemistry. W. W. Norton & Co., New York, p 314
8. Russell CA (1976) The structure of chemistry, Unit 3. Open University Press, Milton Keynes
9. Bensaude-Vincent B, Stengers I (1996) A history of chemistry (translated by D. van Dam). Harvard University Press, Cambridge, MA, p 209–210
10. Barkan DK (1992) A usable past: creating disciplinary space for physical chemistry. In: Nye MJ, Richards JL, Stuewer RH (eds) The invention of physical science: intersections of mathematics, theology and natural philosophy since the seventeenth century. Kluwer, Dordrecht
11. Kauffman GB (1966) Alfred Werner, founder of coordination chemistry. Springer, Berlin
12. Brock WH (1992) The Norton history of chemistry. Norton, New York, p 573–591
13. Werner A (1893) Beitrag zur Konstitution anorganischer Verbindungen. Z anorg chem 3:267–330
14. Brock WH, Jensen KA, Jørgensen CK, Kauffman GB (1982) Searching the Literature To Learn How the Term "Ligand" Became a Part of the Chemical Language. J Chem Inf Comput Sci 22:125–129
15. Hudson J (1992) The history of chemistry. Chapman & Hall, New York, p 187
16. Kauffman GB (1966) Alfred Werner, founder of coordination chemistry. Springer, Berlin, p 64
17. Voss L, Krüss G (1892) Zur Einführung. Z. anorg. chem. 1:1–3
18. Jansen M (2012) Personal communication
19. Kauffman GB (1973) Alfred Werner's research on structural isomerism. Coord Chem Rev 11:161–188 (see footnote, p 174)
20. Wiberg E (1977) Alfred Stock and the renaissance of inorganic chemistry (translated by H. Nöth and R. H. Walter). Pure Appl Chem 49:691–700
21. Werner H (2009) Landmarks in organo-transition metal chemistry: a personal view. Springer, New York, p 85–127
22. Brock WH (1992) The Norton history of chemistry. Norton, New York, p 462–505
23. Hargittai I (2003) Candid science III: more conversations with famous chemists. Imperial College Press, London, p 478
24. Hargittai B, Hargittai I (2005) Candid science V: conversations with famous scientists. Imperial College Press, London, p 353

Chapter 3
The (Re)birth of Inorganic Chemistry

Inorganic chemistry is not general chemistry.... Inorganic chemistry is not general chemistry.

Therald Moeller, *Inorganic Chemistry: An Advanced Textbook* (1952)

Introduction

In 1956 Ronald Nyholm (Fig. 3.1) proclaimed the renaissance of inorganic chemistry in his inaugural address as Professor of Inorganic Chemistry at University College London, where he had recently arrived (from Australia). A paper based on that lecture, which subsequently appeared in the *Journal of Chemical Education* [1], disseminated that thesis to a worldwide audience of chemists, and nearly all references to it have been entirely supportive. But in fact Nyholm's proclamation does *not* hold up as an assessment of historical developments up to 1956; rather, it should be taken more as a hopeful prognostication of things to come—one which, to be sure, *did* come true. One can argue whether Nyholm (and other commentators, both contemporary and subsequent) correctly identified the most important factors and trends that turned out to create the surge of interest he hoped for, a topic I will address in Chap. 5. But there *is* solid evidence for a major transition in the status of inorganic chemistry, beginning around the time of Nyholm's lecture.

In this chapter I will attempt to document that claim: first by looking at recollections from some of the central figures in the nascent field of inorganic chemistry during the 1950s; and then by supporting that anecdotal evidence with analysis of quantitative trends across the years in question. Again, it must be acknowledged that this discussion is primarily concerned with US academia. There are indications that the situation elsewhere, especially in the UK, was much the same, but additional research would be needed to establish that more definitively.

Voices

What was the status of inorganic chemistry in US academia at the time of Nyholm's pronouncement? Recollections of prominent inorganic chemists, concerning programs that are now regarded among the leaders in the field, strongly

J. A. Labinger, *Up from Generality*, SpringerBriefs in History of Chemistry, DOI: 10.1007/978-3-642-40120-6_3, © The Author(s) 2013

Fig. 3.1 Ron Nyholm
(1917–1971), whose leading
role in British inorganic
chemistry was cut short by
his death in a traffic accident.
The photo dates from
sometime in the 1960s
(Courtesy of the Department
of Chemistry, University
College London)

suggest that Nyholm's characterization of earlier days—the quotation that heads
Chap. 2—was still a fairly accurate description in the 1950s. A good place to start
is the University of Illinois at Urbana-Champaign (UIUC), which as we will see (in
Chap. 5) can make a reasonable claim for being the most important locus of the
(forthcoming) renaissance, and for which commentary spanning several decades is
available.

John Bailar (Fig. 3.2), considered by many to be the founding father of
American inorganic chemistry, arrived at UIUC in 1928. He had done his Ph.D. in
organic chemistry (at the University of Michigan, with Moses Gomberg), but he
was hired at UIUC, as an instructor, to teach general chemistry. At the time, he
notes, there was no differentiation between the inorganic and general teaching
faculty, and hardly any instruction in inorganic beyond the freshman program.
Nor was this atypical of American academia; according to Bailar there were only a
handful of universities where it was even possible to get a doctoral degree specializing in inorganic chemistry. He switched his research interests to inorganic
chemistry—in part because he recognized he would have no chance of staying on
as an organic chemist—and started an active program in coordination chemistry;
but there was not much of an inorganic community in the US, and few opportunities to communicate with others [2].

A little over a decade later (1940, to be precise), Fred Basolo (Fig. 3.3) came to
UIUC as a graduate student. Unlike Bailar, he *did* have some prior interest in
inorganic chemistry, having been impressed by one of his undergraduate teachers;
and he decided to work for Bailar, although he seems to have been fully aware,
even at the time, that he was choosing an unpopular field. As he describes it, in

Fig. 3.2 John Bailar
(1904–1991) in 1948
(Courtesy of the University of
Illinois Archives, Faculty,
Staff and Student Portraits,
RS 39/2/26)

1943 "it was believed that beginning general chemistry covered inorganic chemistry, and that no further course on it need be offered nor was there any reason for doing research in the area." Numerically there was still no sign of growing interest at UIUC: only six (out of a couple hundred) graduate students were working in inorganic chemistry [3].

Basolo moved on to Northwestern University (NU) to begin his own academic career. But before we follow him there, we should stay at UIUC to look at the period in which we are most interested, the 1950s. Theodore L. Brown was hired at UIUC in 1956, primarily to teach general chemistry—thus effectively following Bailar's path of nearly thirty years earlier. For his Ph.D., Brown had worked primarily on properties of organolithium compounds at Michigan State University (Michigan State College at the time). His mentor, Max Rogers, was considered a physical chemist, as was Brown during his first few years at UIUC, where he continued his organolithium work. He had thought he would be able to move up the tenure track as a physical chemist, but was disabused of that idea by a senior member of that group, who told Brown that his work would not be considered weighty enough to achieve a tenured position in physical chemistry. On the other hand, said the physical chemist, he might well succeed as an inorganic chemist—the obvious implication being that the latter field was of considerably less importance and hence not subject to the same high standards. Since his interests were already moving in that direction, Brown did change his affiliation, and did enjoy great success, including tenure and a long career at UIUC [4].

Fig. 3.3 Fred Basolo
(1920–2007) in 1964 (Photo
courtesy of Northwestern
University Archives,
Evanston IL)

James P. Collman is another notable inorganic chemist who was at UIUC in the mid-1950s, as a graduate student. Like Basolo, Collman came out of his undergraduate studies (at Nebraska) with an appreciation for inorganic chemistry. He had even done some research there on coordination chemistry, which he believes was extremely rare at that time. Nonetheless, Collman decided it would be better to do his Ph.D. in organic chemistry (although he did try to stay close to the inorganic chemists in the department), and his first appointment (at the University of North Carolina) was as an organic chemist. Only after establishing himself there did he begin his major work in inorganic chemistry, first on coordination chemistry, then (as we shall see in Chap. 5) as an important figure in the central field of organometallic chemistry, and finally in bioinorganic chemistry [5]. It was about the same time, in the mid-to-late 1950s, that UIUC began to increase the presence of inorganic chemistry, hiring figures such as Russell Drago and T. S. Piper, and thus building upon the tradition begun by Bailar to establish itself as a major center in the field.

Back to Basolo: after receiving his Ph.D. in 1943 and working on a war-related research project at Rohm and Haas, he arrived at NU in 1946. Like Bailar and Brown (and many others in these early days), he was hired primarily to teach general chemistry, rather than for his potential research in inorganic chemistry. At the time there was only one other inorganic chemist at NU, and hardly any interest in the field among either existing faculty or new hires. Basolo went to work on recruiting Ralph Pearson, nominally a physical chemist whose program focused on mechanistic studies of organic reactions, arguing that applying the same methods to coordination chemistry would be both novel and essential. Pearson initially demurred, saying "There is no interest in inorganic chemistry, so why should I waste my time?" but eventually came around, leading to a long and fruitful collaboration [3, 6].

By the mid-1950s a viable group of inorganic chemists was beginning to form at NU, under Basolo's leadership; but the field was still well short of high visibility or even parity with other areas of chemistry. Harry Gray came to NU as a graduate student in 1957, with no thoughts of specializing in inorganic chemistry, even though he recalls being fascinated by colors of inorganic materials, such as Prussian Blue, as a child [7]. His undergraduate experience and interests were primarily in organic chemistry. Upon his arrival, though, he was convinced by Basolo that a boom in inorganic chemistry was underway, and that he should get in at that early stage. Following completion of his degree jointly with Basolo and Pearson (and a postdoctoral stint in Copenhagen with Carl Ballhausen, a physical chemist who was one of the pioneers in the theoretical study of coordination complexes), Gray accepted an offer from Columbia University, where he became the first inorganic chemist on the faculty. He had hoped and expected to be the first of a group; but that plan met resistance. The senior faculty at Columbia felt that a small department such as theirs could only do one or two things really well, and hence they should play to their existing strengths in organic and physical chemistry with only a token effort in other fields—a position supported by the Dean. Indeed, not until Gray left Columbia for Caltech—a move for which his senior colleagues' attitude was a significant causal factor—was another inorganic chemist appointed [8].

Many other notable inorganic chemists who got their Ph.D.'s around this time tell similar stories. Starting from a position of unawareness of and/or disinterest in inorganic chemistry, they were ultimately attracted to it, primarily by the intellectual and personal force of a dynamic young faculty member. Richard Eisenberg, one of Gray's first graduate students, was an undergraduate at Columbia, during which time he had essentially *no* exposure to inorganic chemistry. He chose to stay on for his graduate work, liking both organic and physical chemistry—the department's perceived strengths—but was so impressed with Gray's work (and personality) that he decided to become an inorganic chemist [9]—a field in which he has played a major role, most recently as Editor-in-Chief of the ACS journal *Inorganic Chemistry*.

Richard Holm's graduate student experience illustrates both a similar conversion and the nearly ubiquitous attitude of major chemistry departments towards inorganic chemistry of the time. He went to MIT in 1955 leaning towards organic chemistry, which (as at Columbia) was one of the department's two strong areas. Inorganic was not considered to be at all coequal (although since then MIT has become one of the powerhouses of the field); indeed, he found most of the inorganic faculty relatively unimpressive. But he *was* impressed by the youngest member of the inorganic group, Al Cotton,[1] whose studies on sandwich complexes and related topics seemed to him novel and exciting, and the idea of working in a fresh and evolving field was so attractive that he switched over [10, 11].

Likewise, Jack Halpern arrived at the University of Chicago in 1962, primarily as a replacement for Henry Taube (who had just departed for Stanford), and became essentially the sole face of inorganic chemistry at Chicago. (Taube had been thought of mostly as a physical chemist by the Chicago department, although he is certainly now regarded as having been one of the most important figures in inorganic chemistry.) Like Gray at Columbia, Halpern tried to convince his colleagues to augment the presence of inorganic chemistry, particularly by bringing in some synthetic chemists; like Gray at Columbia, he found the atmosphere quite inhospitable to his goals. The departmental attitude changed only after the nearly simultaneous retirement of an entire older generation of faculty members [12].

Fred Hawthorne followed a somewhat different indirect path to inorganic chemistry. He received his Ph.D. in 1953 in organic chemistry, and was hired to his first job (at Rohm and Haas) as an organic chemist. In 1956 he was asked by his supervisor to lead a research effort on boron hydride chemistry for rocket propellant applications, even though at the time he knew nothing about the subject—and not much more about inorganic chemistry in general [13]. Since that (somewhat forced) move, though, he has become one of the leading figures in boron chemistry, as well as inorganic chemistry in general. He moved to academia in the 1960s (first UC Riverside, then UCLA, and finally to Missouri); in 1969 he became Editor-in-Chief of *Inorganic Chemistry*, a position in which he served for a period (32 years) substantially longer than all other editors to date put together.

By the beginning of the 1960s, we can see signs of change: it was more common to find chemists who deliberately sought out opportunities in inorganic chemistry for their professional training. Several examples may be found in a series of interviews with prominent inorganic chemists (carried out and posted online in commemoration of the 50th anniversary of the establishment of the journal *Inorganic Chemistry*). Ken Raymond recalls attending a summer institute of inorganic chemistry that was held at Reed College for several years while he was an undergraduate there, and being introduced to a number of members of the 1950s generation of new inorganic chemists; they so impressed him that he decided to join their ranks [14]. He went to Northwestern for his Ph.D. to study with Fred Basolo, before joining the chemistry faculty at UC Berkeley in 1968, where he has worked on a broad range of topics in bioinorganic chemistry, especially the biological behavior of iron. Steve Lippard turned down offers to

---

[1] F. Albert Cotton (1930-2007) was a graduate student with Geoffrey Wilkinson (see Chap. 4) at Harvard, joining the faculty at MIT in 1955, and moving thence to Texas A&M in 1972. He was one of the most prominent figures in the field of inorganic chemistry throughout the second half of the twentieth century, known for his work in structural chemistry, especially that of species with multiple metal–metal bonds; for the many students and postdocs he trained and sent on to academic positions (see Fig. 5.10); and for the important textbook *Advanced Inorganic Chemistry* that he co-authored with Wilkinson.

attend graduate school at Harvard and Caltech, even though he considered them the top two chemistry departments in the country at the time (1965). Why? He perceived (entirely correctly, as we will see in the next chapter) that they had essentially no presence at all in inorganic chemistry, which was his area of choice coming out of his undergraduate education [15]. Instead he went to MIT, where (like Holm a decade earlier) he joined Cotton's group, and went on to an independent career focusing on bioinorganic chemistry, first at Columbia and then back at MIT.

Numbers

The above anecdotal reports suggest a significant change in attitude towards inorganic chemistry among prospective students, taking place sometime around 1960. Are these merely isolated cases, or do they reflect a more universal trend that can be supported with quantitative evidence? A variety of metrics might serve as indicators of the comparative status of inorganic chemistry:[2] numbers and recognition of faculty in Ph.D.-granting departments; distribution of graduate students by subfield; numbers of papers given at ACS national meetings and published in ACS journals, particularly the *Journal of the American Chemical Society* (*JACS*). As we shall see, all of these data turn out to be completely consistent, both with one another and with the individual stories that signaled that transition in status.

Since 1953 the ACS has issued a biennial "Directory of Graduate Research" (DGR),[3] providing listings of faculty in all US departments of chemistry with graduate programs (later Canadian universities were included as well). Since nearly all faculty entries include their area of specialization, this data set provides a convenient and clearcut measure of how faculty members choose to identify themselves, and to be identified by their peers and prospective students. If we look first at the 1953 edition (which, conveniently, dates from just before the postulated starting point of the transition), we find a total of 90 departments,[4] with a total of 1536 professorial faculty. Of these 203 are identified as inorganic chemists, around 13 %. Even this low number appears to be somewhat exaggerated: fully one in four of them list no publications at all, whereas the fraction of non-publishing faculty in all other specializations is only one in ten. Why such a pronounced difference? Almost surely it reflects the conflation of inorganic and general chemistry which has already been noted: faculty whose primary (or only) departmental role was teaching general chemistry to undergraduates

---

[2] See the section at the end of the book "Notes on quantitative methodology" for explanations of how and from where these data were assembled.

[3] The first edition was called "Facilities, Publications and Doctoral Theses in Chemistry and Chemical Engineering at US Universities, 1953;" the title DGR was used subsequently.

[4] Probably there were a few more that didn't get around to submitting data—indeed there was certainly at least one such, as there is no listing for Harvard in the 1953 edition!

disproportionately tended to call themselves inorganic chemists. Correcting for this effect, we see that the inorganic population of research-active faculty was around 10 % of the total, less than half that of either of the long-established specialties, organic and physical chemistry.

Let us now jump forward 50 years. The 2003 edition of DGR lists 193 Ph.D.-granting chemistry departments in the US (more than double), with 4,800 professorial faculty (more than triple). But that growth is by no means uniform across the range of specializations. There are now 971 inorganic chemists—close to a *five*-fold increase—representing 20 % of the total. Furthermore, there is no longer any significant difference between specializations in non-publishers. The teacher-of-freshman-chemistry/self-described-inorganic chemist has largely gone extinct. Thus the proportion of inorganic chemists within the American professorate roughly doubles over this period, and becomes much closer to (but still probably a little short of) those of organic and physical chemists.

Compiling a similarly detailed set of data for students would be considerably more challenging. One has been published, as part of a general survey of chemistry in the US. The average percentage of graduate students in inorganic chemistry during the period 1924–1935 was reported as around 5.5 %, while the percentage of doctorates awarded in the same subfield for 1967–1975 was 10 % [16]. It is difficult to decide just how to assess the absolute numbers; in particular, why are they so much smaller than faculty proportions? Note that the surveys include a "General *and* Physical" category, so (given the long-standing fuzzy distinctions between inorganic and general chemistry) it is possible that some who might have been appropriately considered inorganic students were classified there instead. Without the raw data we cannot say. In any case, though, the *relative* numbers—an approximate doubling of the relative proportion of inorganic chemists—can be seen in both student and faculty count. The authors of the historical survey concluded that inorganic chemistry exhibited a "strong rise" in interest, whereas that of the other main subfields, physical, organic and analytical, remained quite stable [16].

In addition to quantitative representation, we should also look at *qualitative* measures of status: how the recognition of inorganic chemists by their peers in chemistry as a whole has evolved. How might we measure that? The Nobel Prize, which might first come to mind, constitutes a somewhat problematic gauge, as the award is constrained by some specific guidelines, more restrictive than just recognizing a lifetime body of work. Furthermore, the choice of winner(s) is made by a relatively small body of (Swedish) chemists, who have at times been swayed by personal and even political considerations. The failure to select Mendeleev, who was still alive for the first six years the Nobel was awarded (1901–1906), has been ascribed in part to such factors [17]. Also, of course, Nobel Prizes are not restricted to the US, the focus of this study. Still, the history is (at least) not inconsistent with the proposal that the perception of the field was heightened after the 1950s. There were three early awards that can be considered to recognize inorganic chemistry, either unequivocally (Werner in 1913, for coordination chemistry) or partially (Henri Moissan in 1906, for the discovery of fluorine along with the invention of the electric furnace; Theodore Richards, in 1914, for accurate atomic weight

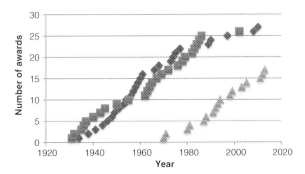

Fig. 3.4 Cumulative number of ACS Awards in Pure Chemistry given to organic (*filled diamond*), physical (*filled square*), and inorganic (*filled triangle*) chemists, 1931–2013 (Other subfields, primarily biochemistry, are not shown)

determination). Of the ninety prizes since then, only seven—less than 10 %—were won or shared by someone who can be clearly identified as an inorganic chemist. But *all* of those latter date from 1973 onwards, giving a representation approaching 20 %.

Restricting ourselves (mostly[5]) to recognition within the US: the ACS sponsors a number of annual awards. Most of them are given for accomplishments in a particular field, and hence not useful for our purpose here; but the ACS Award in Pure Chemistry—one of the most prestigious—does not specify any field.[6] Its defined aim is "To recognize and encourage fundamental research in pure chemistry carried out in North America by young men and women.... Special consideration is given to independence of thought and originality in the research, which must have been carried out in North America" [18]. The only other restriction is that the recipient must be no more than 35 years old at the time the award is given. The frequency of awards to practitioners should thus give us at least some measure of the level of awareness of and respect for the various chemistry subfields (although we are dealing with a statistically much smaller sample compared to the metric of total faculty numbers, of course).

The first Pure Chemistry award was made in 1931, to Linus Pauling. Over the ensuing four decades, nearly all of the awards went to organic and physical chemists, in roughly equal numbers, with a couple of nuclear chemists (and one biochemist) sneaking in. Not until 1970 was an inorganic chemist, Harry Gray, honored with this recognition. But the number of awards to inorganic chemists *since* 1970 is 17, substantially higher than those to either organic (9) or physical (10) chemists, and comprising nearly 40 % of the total! The trend is shown graphically in Fig. 3.4; just as we saw in the faculty and student numbers, the levels of both interest in and visibility of inorganic chemistry in the second half of the twentieth century show a dramatic increase compared to the first half. From

---

[5] Chemists working outside the US are eligible for some of the ACS awards, but not the one in Pure Chemistry.

[6] The Priestley Medal, the top award of the ACS, is also given without respect to field; however, a large fraction of earlier awards (up to 1960 or so) went to industrial chemists and chemical engineers, blurring any easily detectable change over the crucial time period.

Fig. 3.5 Cumulative number of organic (*filled diamond*), physical (*filled square*), and inorganic (*filled triangle*) chemists elected to membership in the National Academy of Sciences, 1940–2012 (Other subfields are not shown)

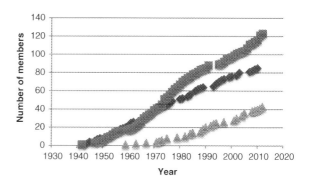

these data we can tentatively date the beginning of this process in the mid-to-late 1950s, as we could expect some lag time before the appearance of strong candidates for an award reserved for young investigators.

Another indicator of recognition may be found in the membership of the National Academy of Sciences (NAS). Those numbers, sorted by subfield, are shown in Fig. 3.5, which (qualitatively, at least) looks very much like Fig. 3.4. There are some differences in quantitative detail, most obviously in the considerable lead of physical over organic chemists. More significantly from our point of view, the *rate* of election of inorganic chemists does not reach approximate parity with the other two subfields until somewhat later, around 1980. A plausible reason for the latter difference is that, unlike ACS awards, only those who are already NAS members can vote on electing new members. That factor might well be expected to result in a sort of autocatalytic effect: voters will be more familiar with, and hence perhaps more likely to support, candidates from their own subfield. Indeed, Basolo attributes the fact that Bailar was never elected to the NAS— a recognition which he surely merited—to the paucity of inorganic chemists who were members and would vote for him [3].[7]

In addition to counting people, we can count papers. Numbers of publications in *JACS*, the most general and most prestigious of the journals published by the ACS, would seem to be the most telling metric. In principle, we could thereby assess *both* activity in *and* regard for the field, as they should be reflected by the number of submissions and the rate of acceptance respectively. However, it is unlikely that sufficient data, parsed by subfield, would be available for a thorough diachronic analysis. One such report suggests that the standing of inorganic chemistry in the early 1950s was not good: the rejection rate for inorganic papers for the period 1952–1955 was noticeably higher than for other fields, although the report also acknowledges the difficulty of assigning submissions to fields [19]. A further potential complication is the introduction of specialist ACS journals during the period under examination, particularly *Inorganic Chemistry* (1962) and

---

[7] A more thorough exploration of these observations could constitute a very interesting sociological study in its own right, but that would go well beyond the scope of this book.

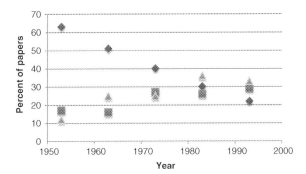

Fig. 3.6 Percent of papers published in *JACS* in organic (*filled diamond*), physical (*filled square*), and inorganic (*filled triangle*) chemistry, 1953–1993

Organometallics (1982), which will likely have perturbed inorganic chemistry submissions to *JACS* relative to those in organic and physical chemistry, for which specialist ACS journals date back considerably further (1936 and 1897 respectively).

Nonetheless, I carried out a spot check of *JACS* data by counting papers in selected issues from one year (the one ending in 3, chosen arbitrarily) for each decade from the 1950s through the 1990s; the results are shown in Fig. 3.6. The dominance of organic chemistry in 1953—over 60 % of the total—is striking, particularly in comparison to physical chemistry. At the time (and for some time thereafter) the most desirable American publication venue for physical chemists was the *Journal of Chemical Physics*, not *JACS* (or the specialist ACS journal, the *Journal of Physical Chemistry*, either). Inorganic chemistry accounts only for around 10 %, a number quite similar to the proportion of inorganic faculty, as we have just seen. But these numbers move dramatically in opposite directions over the years: by 1963 the fraction of inorganic papers already doubled, to around 25 %; and by 1983 it has reached full parity—even superiority—in comparison to both organic and physical chemistry.

It must be recognized that there is a substantial component of subjectivity in this analysis. Papers published in *JACS* have not been identified according to subfield since around 1970, so classifying and counting them involved making my own, necessarily somewhat subjective, classification.[8] Of course, *all* classifications have some degree of subjectivity, but those in the other metrics involve *self*-identification with the subfield, which seems entirely appropriate for a study of how its status is perceived by the larger community.

In that regard, perhaps a better metric may be found in the numbers of papers presented at national ACS meetings. Those *are* almost always explicitly grouped according to the Division to which the author chooses to submit them, and thus may be easily counted, without the need for any subjective decisions on the part of

---

[8] For additional comments on this problematic aspect see Notes on quantitative methodology.

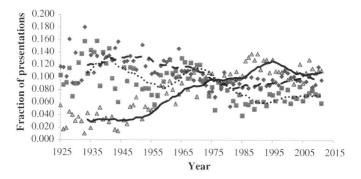

Fig. 3.7 Fraction of presentations at ACS National Meetings made in the Divisions of organic (*filled diamond*), physical (*filled square*) and inorganic (*filled triangle*) chemistry, 1925–2012. The trendlines (organic: *dashed*; physical: *dotted*; inorganic: *solid*) show 10-year moving averages

the analyst.[9] Those data are shown in Fig. 3.7. Note that the *absolute* percentages for all three fields are much lower for presentations at meetings than for publications in *JACS*; that is to be expected, since many papers at meetings are given in Divisions devoted to the more applied fields of chemistry, which generally would not appear in *JACS*. But the trends observed in the two data sets—as well as all the other analyses above—are virtually identical: a very low representation of inorganic chemistry compared to organic and physical chemistry persists through the mid-1950s, and is followed by a rapid and steady increase over the next couple of decades, ending up at quite comparable numbers.

Both the anecdotal and quantitative evidence thus point clearly to a sea-change taking place around 1955–1960. Was it specific to the US? Some have proposed that the status of inorganic chemistry elsewhere was considerably higher before those dates—particularly in Germany but also in the UK, and perhaps even in Canada as well, because of greater transfer of people and interests from the UK there than to the US.[10] I do not have any quantitative data analogous to that cited above, but anecdotal indications suggest that the state of affairs in the UK was in fact not much different from that in the US. Mel Churchill, who did his graduate work at Imperial College in the early 1960s, recalls that when Geoffrey Wilkinson, his co-mentor, returned from Harvard in the mid-1950s he became *only* the fourth chair/full professor of inorganic chemistry in all the UK (the others being at Oxford, Cambridge, and University College; the last was Nyholm, of course) [20].

---

[9] That isn't entirely true: before 1957 there was no separate Division of Inorganic Chemistry, but rather a joint Division of Physical and Inorganic Chemistry, and papers presented therein need to be apportioned between them appropriately. See Notes on quantitative methodology.

[10] Halpern, among others, has offered that suggestion; his own experience (just an isolated example, to be sure) doesn't necessarily support it, since as an undergraduate and graduate student at McGill University in the late 1940s he was exposed to *no* inorganic chemistry at all [12].

Other commentators speak of most universities "virtually ignor[ing] the teaching of inorganic chemistry in the period just after World War II" [21].

Even at ancient Oxford, the history of inorganic chemistry does not go back very far. A committee in 1909 recommended appointing "two University Professors, at £900 a year, one assigned to the subject of organic chemistry and the other to that of inorganic chemistry," but only the organic appointment was actually made. There was a "Dr. Lee's Professorship" of physical and inorganic chemistry, held by Frederick Soddy from 1919 to 1937 (although he had mostly lost interest in chemical research by 1930) and later by Cyril Hinshelwood. The current Inorganic Chemistry Laboratory did not came into independent existence until 1963, when John S. Anderson was named the first Professor of inorganic chemistry [22].

Gordon Stone, who returned from Harvard (like Wilkinson) in 1962, first to Queen Mary College London and then, a year later, to become the (first) Chair of inorganic chemistry at Bristol, attributes the awakening of interest in inorganic chemistry in the UK primarily to Nyholm and Chatt,[11] both of whose research activities gained wide recognition beginning in the 1950s [23]. Although more work would be needed to resolve this question, it does not appear that the status of inorganic chemistry before the 1950s was much higher in the UK than in the US, or that any renaissance there can be dated significantly earlier.

As a sidenote, historian of chemistry William Brock has argued for a major Australian role in awakening interest in inorganic chemistry, pointing to programs there in coordination chemistry during the first half of the twentieth century and, more particularly, a "post-war invasion" of the UK by a number of Australian inorganic chemists,[12] especially Ronald Nyholm [24]. Even though inorganic chemistry has remained a strong presence in Australia, the locus for most of its influence on the US and UK during the period of the renaissance seems clearly to have moved, with Nyholm, to the UK.

In continental Europe, particularly in Germany, we can discern a clearer difference. That was not so much the case at the beginning of the century: one commentator noted that "even at the present day [1899] the full professorships in German universities are almost invariably held by organic chemists, while inorganic chemistry is left to subordinates" [25]. But a strong tradition of research in synthetic inorganic chemistry *did* develop during the early part of the twentieth century, as we saw in Chap. 2. What is striking, in light of what we have seen in this chapter, is how *little* influence that tradition had on American academic research interests. Perhaps that can be attributed—at least in part—to differing attitudes about how research should be approached. I would suggest (obviously

---

[11] Joseph Chatt (1914-1994) spent the earlier part of his career in industry, at Imperial Chemical Industries; while his work there in organo-transition metal chemistry was widely recognized, his impact on the field really burgeoned after he became Chair of inorganic chemistry at the University of Sussex in the early 1960s.

[12] Brock doesn't mention John Anderson, who was not Australian by birth but spent the majority of his earlier career at the University of Melbourne before taking up the inorganic chair at Oxford.

this is a vast overgeneralization) that US researchers have tended to be more impressed by mechanistic, explanatory investigations, whereas the earlier inorganic chemistry research in Germany was mostly phenomenological, centered primarily on synthesis and characterization of new compounds. Fred Basolo tells a story of meeting Hieber at a meeting in 1955. After Hieber's lecture on some new reactions of metal carbonyl complexes, Basolo asked him what he could say about the mechanisms of the reactions; Hieber responded "We do real chemistry in my laboratory, not the philosophy of chemistry" [3, 6].

In a similar vein, Karl Wieghardt, the former director of the Max Planck Institute for Bioinorganic Chemistry in Mülheim, comments that in Germany, mechanistic work was long considered to belong to physical—*not* inorganic—chemistry. He recalls being advised by Nobel laureate E. O. Fischer (who was—perhaps not coincidentally—Hieber's student and successor as Chair of inorganic chemistry at the Technische Hochschule München) that if he wanted a career in inorganic chemistry, he should stop studying mechanisms and go make compounds [26]! The importance of mechanism as a key factor in the emergence of inorganic chemistry as an intellectually respectable field in the US will be seen in Chap. 5.

To sum up: Nyholm's proclamation of a renaissance in inorganic chemistry in 1956 was premature but prescient, as was Moeller's claim (see the epigram that heads this chapter) that inorganic chemistry had (finally?) managed to distinguish itself from general chemistry [27]. Only the first stirrings of those breakthroughs show up in the data and reminiscences from the early to mid-1950s, but by the 1960s they are already clearly visible, leading over the next couple of decades to full parity. In the next chapter I digress a little to focus on the history of this evolution at one institution and the personal factors that strongly influenced it, before returning to the broader scene in Chap. 5, where I examine the agents responsible for the change.

References

1. Nyholm RS (1957) The renaissance of inorganic chemistry. J Chem Educ 34: 166–169
2. Bailar JC (1987) John C. Bailar, Jr., interview by Theodore L. Brown at University of Illinois, Urbana, Urbana, Illinois, 28 May & 17 June 1987 (Philadelphia: Chemical Heritage Foundation, Oral History Transcript # 0073)
3. Basolo F (2002) From Coello to inorganic chemistry: a lifetime of reactions. Kluwer, New York, p xii
4. Brown TL (2012) Personal communication
5. Collman JP (2012) Personal communication
6. Basolo F (2002) Fred Basolo interview by Arnold Thackray and Arthur Daemmrich at Northwestern University, Evanston, Illinois, 27 September 2002 (Philadelphia: Chemical Heritage Foundation, Oral History Transcript # 0264)
7. Gray HB (2011) Voices of inorganic chemistry (interview with Richard Eisenberg) on website http://pubs.acs.org/page/inocaj/multimedia/voices.html
8. Gray HB (2012) Personal communication

9. Eisenberg R (2012) Personal communication
10. Holm RH (2012) Personal communication
11. Holm RH (2011) Voices of inorganic chemistry (interview with Richard Eisenberg) on website http://pubs.acs.org/page/inocaj/multimedia/voices.html
12. Halpern J (2012) Personal communication
13. Hawthorne MF (2011) Voices of inorganic chemistry (interview with Richard Eisenberg) on website http://pubs.acs.org/page/inocaj/multimedia/voices.html
14. Raymond KN (2011) Voices of inorganic chemistry (interview with Richard Eisenberg) on website http://pubs.acs.org/page/inocaj/multimedia/voices.html
15. Lippard SJ (2011) Voices of inorganic chemistry (interview with Richard Eisenberg) on website http://pubs.acs.org/page/inocaj/multimedia/voices.html
16. Thackray T, Sturchio JL, Carroll PT, Bud R (1985) Chemistry in America 1876-1976: historical indicators. Reidel, Dordrecht, p 186–187
17. Lagerkvist U (2012) The Periodic Table and a missed Nobel Prize. World Scientific, Singapore, p 100–114
18. From the ACS website, http://webapps.acs.org/findawards/detail.jsp?ContentId=CTP_004546. Accessed 17 January 2013
19. Larsen EM (1957) The crisis in inorganic chemistry. J Chem Ed 34:427–429
20. Churchill MR (2012) Personal communication
21. Lord Lewis of Newnham, Johnson BFG (1997) Cyril Clifford Addison: 28 November 1913-1 April 1994. Biographical Memoirs of Fellows of the Royal Society 43:2–12
22. http://www.chem.ox.ac.uk/history/. Accessed 28 January 2013
23. Stone FGA (1993) Leaving no stone unturned: pathways in organometallic chemistry. ACS, Washington
24. Brock WH (1992) The Norton history of chemistry. Norton, New York, p 603–618
25. Stokes HN (1899) The revival of inorganic chemistry. Science 9: 601–615
26. Wieghardt K (2011) Voices of inorganic chemistry (interview with Richard Eisenberg) on website http://pubs.acs.org/page/inocaj/multimedia/voices.html
27. Moeller T (1952) Inorganic chemistry: an advanced textbook. Wiley, New York, p 3

Chapter 4
The Personal Factor: Donald Yost and Inorganic Chemistry at Caltech

> *"Him that we do not name"*
> J.R.R. Tolkien, referring to Sauron,
> in *The Fellowship of the Ring*
> *"He-Who-Must-Not-Be-Named"*
> J. K. Rowling, referring to
> Voldemort, in *Harry Potter
> and the Chamber of Secrets*
> *"The unnamed person"*
> D. M. Yost, referring to Linus Pauling,
> in a letter to Harold Johnston

The preceding chapter emphasized universal trends in the status in inorganic chemistry in the US, but individual behaviors and local variations—how the story played out in particular institutions—are also of considerable interest. In many cases these have a strong personal component: the attitudes of a powerful faculty member or group of members, or even the relationship between members, may strongly influence the way things go. We have already seen how Werner's management style may have dampened his protégés' enthusiasm for continuing in the field (Chap. 2), and, conversely, that strong personalities were in large part responsible for recruiting students to inorganic chemistry who had no prior interest in the field, but came to play major roles in its growth. As another example, it has been suggested that organic chemistry was relatively weak at UC Berkeley for a long time, because G. N. Lewis felt that only physical chemistry was important [1].

I have already noted (Chap. 1) that it is hard to see any signs of the general increase in respect for inorganic chemistry following the 1950s at Harvard. That is more than a little ironic, since many of the most important early developments in organometallic chemistry, which as we shall see in Chap. 5 were central to the renaissance of inorganic chemistry, took place there. But the ways of the Harvard chemistry department—particularly a strong tradition of *not* promoting from within—made it much harder to change the faculty's rather low opinion of the field.

Harvard apparently *did* periodically feel that inorganic chemistry merited increased attention—a 1946 letter from then Chair George Forbes states "I am anxious to increase departmental activity in the field" [2]—but that sentiment translated into very little visible presence, aside from Rochow,[1] until Geoffrey

---

[1] Eugene G. Rochow (1909–2002) was a leader in the field of silicon chemistry, first at General Electric and then, from 1948–1970, as (the only) tenured professor of inorganic chemistry at Harvard.

J. A. Labinger, *Up from Generality*, SpringerBriefs in History of Chemistry, DOI: 10.1007/978-3-642-40120-6_4, © The Author(s) 2013

Wilkinson (Fig. 4.1) came to Harvard in 1951. He was actually hired as a nuclear chemist—a field that had gained considerable status during and after World War II—but switched his entire program to organometallic chemistry following his (jointly with R. B. Woodward) correct structural interpretation of the novel, recently discovered molecule ferrocene (see Chap. 5). Nonetheless, despite his seminal contributions to what was obviously becoming a hot topic, the department denied him tenure in 1956, as part of what Gordon Stone called "…one of its frequent exercises of dispensing with the services of several nontenured teaching staff…" and "a general reluctance on the part of most tenured faculty to have any respect for synthetic inorganic chemistry" [3]. Years later, after Wilkinson shared the 1973 Nobel Prize (with E. O. Fischer) for his work on ferrocene, Woodward reportedly remarked that he *still* thought Harvard had made the right decision in letting him go [4].

A number of inorganic chemists appeared on the Harvard faculty during the years following Wilkinson's departure. Gordon Stone, who was already present as a postdoc (with Rochow), was given a faculty position and kept alive Harvard's substantial standing, started by Wilkinson, in organotransition metal chemistry. Subsequent appointees included Dick Holm, Alan Davison, Mel Churchill, John Osborn (my graduate mentor), Stan Wreford, John Cooper, Steve Cooper, David Hoffman, and Andy Barron. Most of them (like Wilkinson) were British; most of them (like Wilkinson) specialized in organometallic chemistry; *all* of them (like Wilkinson) went through the same revolving door of tenure denial [5]. Not until 1980, when Holm was re-hired (from Stanford), was there once again a tenured inorganic chemist on the Harvard faculty; and although Holm was told at the time of his return that he would have the opportunity to build up a group in the field, that didn't happen to any significant extent until many years later [6].

For the rest of this chapter I will take a detailed look at my own institution, Caltech. Today, and for the last several decades, its inorganic group incontestably ranks among the top few in US academia. Equally incontestably, that was *not* the case before the 1960s; and there are clear indications that interpersonal interactions—or perhaps more precisely, a lack thereof—had much to do with that state

Fig. 4.1 Geoffrey Wilkinson showing off some of his chemistry, ca. 1985 (Photo courtesy of the Department of Chemistry, Imperial College London)

Fig. 4.2 Linus Pauling in his laboratory, ca. 1930s (Photo courtesy of the Archives, California Institute of Technology)

of affairs. In the course of this examination we will meet a character (in every sense of the word!) who was the main representative of inorganic chemistry at Caltech up through the 1950s, although he is not much remembered today.

We may start with a document written by Linus Pauling in 1944, entitled "The Division of Chemistry and Chemical Engineering at the California Institute of Technology: Its Present State and Future Prospects." Pauling (Fig. 4.2) was then Chair of that Division, a position he held from 1937 to 1958. In this memo he set out to assess the strengths and needs of the various subfields of chemistry, opining in the first section that "Professor Arthur A. Noyes developed...a Department of Chemistry which was very strong in the fields of physical and inorganic chemistry. The present members of the staff...are doing and can be expected to continue to do a thoroughly satisfactory job of teaching general, analytical, physical, inorganic and colloid chemistry...[and] will without doubt continue to carry on research which will uphold the excellent reputation that the California Institute of Technology has made in these fields. I do not believe that any expansion of the staff...is needed in [these] fields...." [7].

That sounds straightforward enough, but dissonances appear on closer examination. First, it is far from clear that Noyes created—or even intended to create—a department "strong...in inorganic chemistry." Noyes, one of the "triumvirate" (Fig. 4.3) that essentially created Caltech in its modern form, was one of the leading figures of physical chemistry in the US. He had established a major program at MIT—the Research Laboratory of Physical Chemistry—and his principal aim for his new home was to achieve comparable excellence in physical chemistry at Caltech [8]. Starting around 1912, Caltech waged a strenuous and prolonged struggle to entice him away from MIT and bring a similar degree of prestige to chemistry at Caltech, an effort that finally succeeded in 1919 [9]. But there is no evidence of any significant presence of, or interest in, inorganic chemistry during the early years.

The Caltech catalog for 1920 lists eight faculty members, of whom only one is designated as an inorganic chemist: Roscoe Dickinson, who was an instructor, not

Fig. 4.3 The Caltech
"triumvirate"—physicist
Robert A. Millikan (*center*),
astronomer George Ellery
Hale (*right*), and chemist
Arthur A. Noyes (*left*),
affectionately known as
"tinker, thinker, and stinker."
The 1929 portrait by
Seymour Thomas is
prominently displayed in the
Caltech Athenaeum, and was
photographed by Robert Paz
(Courtesy of the Archives,
California Institute of
Technology)

part of the professorial faculty. By 1925 Dickinson had been promoted to
"Research Associate in Chemistry," and *none* of the faculty (now around a dozen,
several of them also Research Associates, including newly-appointed Pauling)
identified their field as inorganic. The course listings are likewise telling: in 1920
there was a two-term course, Chem 301/302, called "Inorganic Chemistry;" but
the course description makes it clear that it was the general chemistry course taken
by all Caltech students in their freshman year. By 1925 the same course had been
redesignated as Chem 1abc,[2] "Chemistry." In neither catalog does *any* more
advanced course appear with any reference to inorganic chemistry in either the
title or description [10]. At least for these first years after Noyes' arrival, inorganic
chemistry was pretty much entirely equated to general chemistry—an attitude by
no means unique to Caltech, as I have argued earlier.

Even more disconcerting is Pauling's summary of staff and courses in the 1944
memo, reproduced in Fig. 4.4. Where are the "normal" staff and courses in
inorganic chemistry that the title would certainly lead us to expect? Either Pauling
still did not care to distinguish inorganic from general chemistry (but he *does* do so
in the passage quoted above), or there is a rather serious omission here! In fact the
latter is the case: there *were* advanced courses in inorganic chemistry, taught by a
member of the staff who *was* designated as an inorganic chemist, but was completely left out of Pauling's listing. That was Donald M. Yost, to whose career and
(shall we say) idiosyncratic personality we now turn—including in particular the
relationship between Yost and Pauling, which may well have played a significant

---

[2] That is, a course extending over all three terms of the year.

```
Normal staff in Physical, Inorganic, Analytical, and Colloid Chemistry:

General chemistry, Ch 1a, b                        Professor Pauling
                                                   Professor Bell
                                                   Assistants

Analytical chemistry, Ch 11, Ch 12a, b            Professor Swift
                                                   Assistants

Physical chemistry, Ch 21a,b; Ch 22; Ch 122a, b   Professor Bates
                    Ch 23a,b                        Professor Dickinson
                    Ch 24a,b                        Dr. Corey

Phys. chem. laboratory and
     colloid chemistry, Ch 26a,b; Ch 29; Ch 129   Professor Badger
```

Fig. 4.4 Excerpt from Pauling's 1944 memo [7]

role in the rather unimposing stature of inorganic chemistry at Caltech during this time.

Don Yost grew up in rural Idaho, and studied chemistry as an undergraduate at UC Berkeley. After graduating in 1923, he first went to the University of Utah for a year, but then transferred to Caltech to work with Noyes for his Ph.D. In 1926—within a year of Pauling's initial appointment—he joined the faculty: first as research fellow, then instructor, and on up the professorial ranks [11, 12]. Although he was considered (and considered himself) to be an inorganic chemist from the beginning, in the first few Caltech catalogs he is listed simply as Assistant Professor of Chemistry (as were several other faculty members). Not until 1936 was he explicitly called Assistant Professor of Inorganic Chemistry, and he remained the *only* professorial faculty member so designated, out of more than twenty total (Fig. 4.5), right up until his retirement in the early 1960s [10].

Yost's arrival brought advanced courses in inorganic chemistry into the curriculum for the first time. The 1930 catalog lists: "Ch. 13a, b. Inorganic Chemistry. The chemical and physical properties of the elements are discussed with reference to the periodic system and from the view-points of atomic structure and radiation-effects. Such topics as coordination compounds, the liquid ammonia system, the compounds of nitrogen, the halides, and selected groups of metals are taken up in some detail." There was also an accompanying lab course, and a graduate-level course entitled "Inorganic Chemistry (Seminar)," which was listed as taught jointly by Yost and Noyes, and which focused on physical aspects of inorganic chemistry [10]. So we may ask again: how, in the (obviously important) memo cited above, could Pauling have thought to omit Yost? By good fortune, from around 1940 on Yost began to save all his papers,[3] and they have been preserved in the Caltech archives, making it possible to trace the rather fractious relationship between Yost and his colleagues, especially Pauling. It may be too fanciful—but

---

[3] I really do mean *all*: clippings from newspapers, receipts, used train tickets, etc. But the files primarily consist of a rich trove of correspondence and documents.

Fig. 4.5 Faculty of the Division of Chemistry and Chemical Engineering at Caltech, 1950. Don Yost is at the left end of the third row; Linus Pauling is third from left in the second row (Photo courtesy of the Archives, California Institute of Technology)

maybe not!—to suppose that Pauling felt justified in excluding Yost from a listing under the heading of "*normal* staff."

In addition to teaching inorganic chemistry, Yost carried out an active research program, at least for the first 20–25 years of his career at Caltech. Most of his work could be best described as physical inorganic chemistry, emphasizing determination of physical and thermochemical properties, spectroscopy (especially Raman), and—perhaps most forward-looking—mechanistic studies. Of his first four papers (which, as reprints, made up the whole of his doctoral thesis), all published in 1926, three (all published in *JACS*) dealt with kinetics and catalysis of inorganic reactions, as did about 20–25 % of the 100+ technical papers he published subsequently, up to the very last one in 1959 (but more on that later). Yost was nationally recognized as one of the outstanding inorganic chemists of this period—certainly at least up through the end of the war, and to some extent for a few years thereafter—and was elected to the National Academy of Sciences in 1944. He published two books, with graduate student co-authors, on the chemistry of the elements from main groups V and VI [13] and of the rare earths [14], both of which were very well received [11]. Joel Hildebrand said of the former "If ever there was a perfect scientific book this is it" [15].

Yost and his work are little remembered today; I have not found a single inorganic chemist whose career began after 1965 or so, and who had no direct Caltech connection, who has even heard of him! But one of the projects he undertook certainly would have ensured his place in the chemistry pantheon, if it had been successful. In 1933 he and a student tried to demonstrate that the "inert"

Fig. 4.6 Don Yost in 1948, tinkering with a piece of apparatus (Photo courtesy of the Archives, California Institute of Technology)

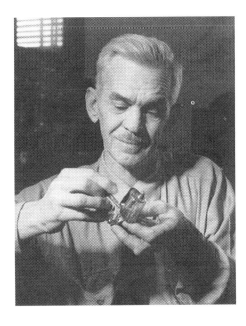

gas xenon might not actually be inert. They subjected mixtures of Xe and F_2 in quartz vessels to photolysis and to an electrical discharge, but saw no evidence of any product in either case (although they did observe that *something* appeared to have attacked the surface of the quartz in the latter case). An account of their experiment was published in *JACS* (a rare example of getting a negative result into that august journal!), ending with the comment that "It cannot be said that definite evidence for compound formation was found. It does not follow, of course, that xenon fluoride is incapable of existing" [16]. Indeed, XeF_2 was synthesized—by exactly Yost's method—nearly 30 years later [17], after Neil Bartlett had broken the first ground in rare gas chemistry.

Why did Yost's attempt fail? It was undeniably an extremely difficult experiment, especially at that time—it could easily have been ruined by the presence of trace amounts of moisture, or less-than-excellent quality materials—although Yost's reputation as a careful experimenter was high (Fig. 4.6). One possible speculation—although it is just that—concerns the degree of his commitment to the project. Pauling had previously predicted that xenon fluorides might exist: it has been reported that Yost carried out the experiment at Pauling's instigation [18], and that Pauling obtained a sample of xenon and gave it to Yost to try the experiment [19]. A suggestion was even offered that Yost's dislike for Pauling might have had something to do with the failure [11]. Unfortunately the Yost files in the Caltech archives do not go back that far, so no first-hand commentary is available. A memoir, written much later, that Yost contributed to a monograph on noble gas chemistry discusses experimental difficulties but not motivations [20]. It is far from clear that their mutual antipathy was *that* severe at such an early date, although it did become so later. Indeed, Pauling felt that at least part of the bad relationship was a *consequence* of what he variously called this "fiasco" [21] or Yost's having "distinctly missed the boat" [19]. It does seem extremely far-fetched to suppose that Yost would have *deliberately* chosen to miss out on what he certainly would have recognized as a chance at immortality, just to spite Pauling. Perhaps it is less unlikely, though, that he didn't give it his all.

Yost was significantly involved in the war effort, seeking (and obtaining) a Naval Reserve commission (he had enlisted in the Navy during World War I), and participating in war-related research, including work on chemical warfare [22]. He was actively recruited for the Manhattan Project and apparently had every inten-tion of moving to Los Alamos [23], but in 1943 he became seriously ill, with an infection of the jawbone, that kept him from going there—or doing much of anything at all—for many months. The disease indeed could well have been fatal, but for his being treated with penicillin, which at the time was newly developed and very difficult to obtain [11]. It is possible that Pauling played a role in pulling the strings needed to get a supply for Yost; even that Yost was a real ingrate for not being sufficiently appreciative. Caltech biologist James Bonner later recalled "Don Yost owed his life to Linus who…arranged for him to get penicillin, which was generally unavailable…and he survived to 97 (*sic*; in fact Yost lived to 83)!" [24]. Again, though, there is nothing explicit on this in the Yost archives.

After the war Yost returned to Caltech and resumed his research program. He continued to be highly regarded by the community at large; his archival files include correspondence about job offers from a number of government labs (Argonne, Brookhaven, Los Alamos) and universities (Chicago, Syracuse, Bran-deis, UC Riverside). Several of the latter asked him to consider coming to chair their chemistry department, a move that would have surely been catastrophic for all concerned, given his iconoclastic attitude towards institutions (as we shall see shortly); he wisely declined to pursue those opportunities. It appears that he *did* accept an offer from Argonne, only to change his mind shortly thereafter [25]. A 1949 letter from Henry Taube asks to visit Yost and pick his brain on kinetics of substitution reactions [26]—a considerable compliment from someone of Taube's stature.

Yost's interest in research seems to have fallen off considerably from this point on. There are a couple dozen post-war publications, the last being the afore-mentioned 1959 mechanistic study on hypophosphorous acid [27]. However, it is clear from an epistolary exchange with his co-author [28] that the experimental work had been completed at least 6 years earlier, and the completion of the final manuscript was a long and unhurried process. By around 1953 he had completely stopped taking on graduate students [29]. His distaste for the institutional aspects of carrying out a research program may well have been a partial cause, as indicated in a later letter [30]:

> There isn't much inorganic chemistry being done here at this Institute these days. The reasons are various, but some of them were due to struggles (competitions). First there was the struggle between professors and bookkeepers; the bookkeepers won. Then there was

the struggle about admitting girls to the graduate school; the girls won.[4] Next came the struggle between organic-biochemistry and physical-inorganic chemistry; the organikers won. Along with these struggles came one between professors and politicians; the politicians have almost won. Being of a scholarly turn of mind I side-stepped the struggles by turning to mathematics.

It *is* arguably true that in terms of graduate student progeny, Yost's most important legacy dates from the 1950s, when two major figures in the development of NMR—John Waugh and James Shoolery—did their degrees with him. Waugh actually did an NMR project for his Ph.D., using the technique (on a machine he built himself) to determine the structure of bifluoride, FHF^-, and went on to a long career at MIT, where he made many crucial contributions, especially in the development of solid-state NMR [31]. Shoolery worked on microwave spectroscopy, but caught an interest in NMR from Waugh, and on completion of his degree went to Varian, where he also had a long career, playing a central role in making NMR the useful and ubiquitous tool for chemistry that it became [32].

However, the memoirs of both of those NMR giants make it clear that Yost actually contributed very little scientifically to their graduate work, acting primarily as a supportive, interested, but mostly hands-off mentor. Waugh: "I became Yost's NMR student.... However neither Yost nor anyone else at Caltech had any experience with NMR. I had to learn the subject myself by reading the relatively few existing papers" [31]. Likewise Shoolery, who notes that Yost knew "absolutely nothing about microwave spectroscopy" but was happy to let him use his space and equipment: "[he] realized that being serious hadn't got him anywhere in life, and so he was very relaxed about most things, including my work. Dr. Yost probably wouldn't have noticed my progress had I never gone to his office" [32].

The post-war Yost seems to have been equally unenthusiastic about teaching inorganic chemistry, an attitude that may well have gone back much further. A striking apologia appears in a 1959 letter: "For one who can get up a lot of steam about scholarly subject matter but who does not like to teach classes, I get into the most remarkable predicaments of anyone I know. I'm not a good teacher, and I often marvel about how I happened to get into such a profession. But its (*sic*) too late to do anything about it now" [33]. Fred Anson[5] recalls Yost's teaching style: he never announced what he was going to talk about, much of which was on topics of dubious relevance to inorganic chemistry, but just got up in front of the class, lit up the first of an endless sequence of cigarettes, and mumbled through a lecture [34]. Waugh also mentions Yost's "uninspiring" lectures [12]. During his

---

[4] There were no women graduate students in chemistry at Caltech until 1953, when Jack Roberts urged a change in policy so he could bring one of his students from MIT. Divisional meeting minutes show a vote of 9-4 in favor; Yost was one of those opposed.

[5] F. C. Anson was an undergraduate at Caltech, went to Harvard for his Ph.D., and returned to join the Caltech faculty in 1957, quickly becoming a leading figure in the field of electrochemistry.

last few years at Caltech his courses on "inorganic chemistry" were reputed to have been almost entirely given over to Boolean algebra [4].[6]

Yost's attitude (and, occasionally, behavior) towards most institutions may be classified as somewhere between idiosyncratic and curmudgeonly—probably tending towards the latter end of the spectrum. His opinion of the ACS provides an excellent example: initially a member, he dropped out in 1938, and responded over the years to regular invitations to rejoin with comments such as "The ACS is interested in petrol chemistry, and that is that;" "It is a source of regret to me that the ACS has so little interest in pure science, and that it is thirty to 40 years behind the times in general;" and "As everyone knows, the ACS, as a purely scientific society, is long obsolete; its (sic) an engineering society" [35]. Responding to a solicitation for nominees for the National Medal of Science, he proposed (in addition to some legitimate candidates) Robert Frost, Carl Sandberg, Henry Miller, Winston Churchill and Henry Ford II [36]! He rarely sought (or received) governmental funding for his work; a number of comments in his correspondence proclaim a philosophical objection to the entire practice (although others suggest that his low anticipation of success might have been a telling motive as well).

As for others in his field: often he was highly complimentary, but he could be (unfairly) dismissive too. In a 1964 letter he comments "There has been a vast amount of work done on coordination compounds, but when you come right down to it none of it goes much further than Werner left the subject" [37]. (That opinion didn't stop him from signing a contract with Wiley for a book on coordination compounds [38], but like the planned math text, it was never completed).

The foregoing (not to mention some of the forthcoming) might lead one to describe Yost as a crank, and perhaps even a failure—but that would be a mistake. As we saw above, he made important contributions to the field, particularly in the pre-war segment of his career, for which he achieved national recognition. He always had a large number of fervent admirers, especially among his former students, many of whom remained devoted friends; his biographers stress that point [11, 12], and it is eminently clear in much of the correspondence found in his files. The latter contain many appreciative responses to his crusty style, which particularly characterized his many book reviews. The editor of *Science* wrote to Yost, concerning the latter: "All I can say is, any time that you care to send us a paper of the general or survey type...you may be sure that it will be welcomed" [39].

It is also quite likely that some of what we have seen[7] of his cantankerousness was, at least partially, for show. In a 1959 response to an announcement of a forthcoming Gordon Research Conference on inorganic chemistry, he says: "Of course I realize that we in the west are terribly provincial, and a priori we count all those living east of Dodge City as gringos, but at the same time I do feel a bit

---

[6] There are several mentions in the archives of his substantial progress towards writing a math text, but it was apparently never completed.

[7] Also, of course, one is tempted to select the most colorful incidents for their entertainment value—a temptation I have only occasionally resisted.

embarrassed in having to ask you directly just what in blazes is the Gordon Research Conference?" [40]. Of course Yost must have been completely familiar with the Gordon Conferences; there are earlier invitations in the file, and even a letter *from* him to one organizer, offering suggestions for possible speakers [41]! There are many other such indications that Yost deliberately cultivated such an iconoclastic and ruggedly individualistic persona.

Nonetheless, it is beyond question that his attitude and behavior towards his Caltech colleagues were *not* just for show, and had real, significant consequences. His relations with Pauling were particularly problematic, despite (or perhaps because of?) their rather similar personal backgrounds. Like Yost, Pauling's roots were in the still-somewhat-Wild West (rural Oregon rather than Idaho); one biographer describes him as a "brash intellectual frontiersman" [42]. To be sure, their personalities developed in very different directions—Pauling was always a large and visible presence in the public eye, while Yost preferred to remain much more private—but neither was much inclined to make accommodations for those who disagreed with them. In particular, Pauling did not get along well with either Noyes or Millikan, two of the three founding giants of Caltech [43].

Religion and politics may well have been further sources of interpersonal friction. Pauling was strongly antireligious, one of the reasons for his antagonism towards Millikan [44]; in contrast, there are many papers in Yost's archival files that show him to have been a committed Catholic, although the strength of his commitment was probably diluted by his dislike of institutions in general. And Pauling's left-wing political stances, from the 1940s on, were always on prominent display, whereas Yost was, at least originally, a staunch Republican.

To digress for a moment: the political arena provides another illustration of how Yost's anti-establishment tendencies trumped institutional loyalties. In 1948, asked to support Edward Condon (whose loyalty was being investigated by the House Un-American Activities Committee), he commented that HUAC has "the right to investigate the political activities and loyalties of any and all public servants and other residents of this country" [45]—a surprisingly passive position for one so skeptical of authority, government or otherwise. But he soon changed his mind: a lengthy letter to a former student expresses considerable outrage about the J. Robert Oppenheimer hearings [46]. Reportedly[8] Yost was visited one day by an FBI agent asking for "information" about Oppenheimer. Yost said "Sit down, son," and asked for the agent's credentials. He then took a gigantic magnifying glass out of his desk drawer, spent several minutes looking back and forth from the ID to the person, and finally handed the ID back saying "This doesn't look like you, son." The agent stammered a few words, and fled [4]. By 1972, Yost's Republican leanings had so far eroded that he voted for McGovern [47]!

---

[8] I've heard this story independently from several people, each of whom had heard it at second or third hand. But of course that doesn't guarantee its authenticity, as the original source could only have been Yost himself.

Returning to Yost's relationship with the CCE Division: Jack Roberts recalls that when he came to Caltech in the early 1950s, Pauling took him aside and explained how Divisional governance worked. He didn't like issues to be settled on closely divided votes, Pauling told Roberts, and always sought a substantial consensus; but there would *never* be a unanimous vote, because Yost could be counted on to vote against everything [48]. Almost certainly Yost's personal animosity towards Pauling was a major contributor to that attitude. Yost steadily withdrew from any involvement in the CCE Division while Pauling remained chair. In a 1947 memo he asks to be relieved of "all future responsibility for the qualifying examination in inorganic chemistry" [49]. Correspondence with several people reveals Yost's disrespect for Pauling as Chair; at its most extreme: "the unnamed person...is extremely naïve and, as two Deans put it, pathologically single tracked and nuts" [50].

That last opinion seems to have been held mutually. In a 1956 memo Pauling reminds Yost that at an earlier meeting that year, concerning Yost's reported refusal to work with or even talk to a prospective graduate student, "I assume that this statement made by the man is true, because I know that you refused to talk with me, when I came to your office, that you put your hands over your ears during part of the time that I was in the office, and that you did not answer any of the questions that I asked you" [51]. (As bizarre as that behavior sounds, the fact that Yost kept this memo for posterity may be equally indicative of his attitude!)

Yost seems to have had a more hopeful outlook for the new regime; minutes of Divisional meetings [52] show Yost attending fairly regularly through the 1940s, then tapering off and ceasing altogether in the mid-1950s, but resuming in 1958— just when Pauling left Caltech and the Divisional chairmanship. However, his attendance soon ceased again. His annual "reports" to the Division chair, required of all faculty members, are characteristic: a typical example reads, in its entirety [53]:

> Research: Yes
> Financial Support: Newmont
> Publications: Yes
> News: Quicquid agunt homines nostri est farrago libelli.
> Respectfully submitted, Don M. Yost.

Subsequent ones, even after Swift replaced Pauling, were identical (except for varying the Latin phrase).

In light of all the above, it is unsurprising that Pauling showed no inclination to raise the stature of inorganic chemistry at Caltech; and if Yost had any such inclination (which seems most unlikely), he had no clout at all within the Division. According to Roberts, Pauling considered himself enough of an inorganic chemist[9] that no others were really needed [48]. When Roberts first proposed adding

---

[9] In fairness, that opinion was not *entirely* unjustified: Pauling, who often called himself a "structural chemist," had surrounded himself with a group of crystallographers whose main interests were in inorganic chemistry [4].

Fig. 4.7 Harry Gray delivering a lecture, shortly after moving from Columbia to Caltech in the 1960s (Photo courtesy of Harry Gray)

someone in inorganic chemistry, recommending Henry Taube as a good candidate, Pauling—who had complete control of hiring at the time—was completely uninterested [34].

Finally, after Pauling's departure, Caltech began to pay more serious attention. By the early 1960s, as seen above, there was effectively *no* inorganic chemistry at Caltech, either in research or instruction. Roberts and others made inquiries and identified Harry Gray (Fig. 4.7), then at Columbia, as a prime target [48]. Gray wasn't really anxious to leave New York, but was excited by the possibility of leading a strong program in inorganic chemistry—Roberts told him he could expect to assemble a group of five or six faculty members—whereas there were no prospects for growth at Columbia, as recounted in the previous chapter. Gray came for a short period as a Visiting Professor in 1965, moved for good the following year, and immediately set about building up the field. The first couple of hires didn't work out—one left to join the Church of Scientology [4]!—but John Bercaw, who arrived in the early 1970s, is still at Caltech today (as is Gray). Over the next few decades the inorganic chemistry group gradually increased to the present day count of six, about one-fifth of the total chemistry faculty, equal to the number of organic chemists at Caltech. The strength of the inorganic program is also beyond question: it is routinely ranked among the top two or three in the country, and at least on a par with the overall outstanding reputation of chemistry at Caltech.[10]

In retrospect, then, it is plain to see how inorganic chemistry at Caltech has been crucially affected by personalities. Before World War II, the Caltech inorganic program was quite limited in scope—one person!—but that was not particularly

---

[10] One striking statistic supporting that claim: fully five of the six have received the ACS Award in Pure Chemistry, a record that is surely unmatched by any other subgroup at any other university.

atypical of the national status of the field. Soon after the war inorganic chemistry virtually disappeared from Caltech, until it was reborn in the 1960s to take its leading place among US university programs. And much—probably most—of that evolution was intimately linked to the personal characters and beliefs of two people—Pauling and Yost—and their fractious interactions. Not until both were out of the picture did it become possible to establish a program in inorganic chemistry that matched the stature of the Division and the Institute as a whole.

References

1. Taube H, Gortler L (1986) Henry Taube, interview by Leon Gortler on the way to Grand Central Terminal, New York, New York, 19 March 1986 (Philadelphia: Chemical Heritage Foundation, Oral History Transcript #0298)
2. Forbes G (1946) Letter to DM Yost, 2/20/46, from the Yost file, #6.6, Caltech archives
3. Stone FGA (1993) Leaving no stone unturned: pathways in organometallic chemistry. ACS, Washington
4. Gray HB (2012) Personal communication
5. Churchill MR (2012) Personal communication
6. Holm RH (2012) Personal communication
7. Pauling L (1944) From the CCE Division file, #1.12, Caltech archives
8. Servos JW (1990) Physical chemistry from Ostwald to Pauling: the making of a science in America. Princeton University Press, Princeton NJ, p 100–105
9. Goodstein JR (1991) Millikan's school: a history of the California Institute of Technology. Norton, New York, p 51–63
10. Catalogs from the Caltech archives
11. Cole T (1977) Don M. Yost, 1893-1977: a tribute by Terry Cole. Engineering and Science 41:28–29
12. Waugh JS (1993) Don Merlin Lee Yost (1893–1977): a biographical memoir. From the National Academy of Sciences website, http://www.nasonline.org/member-directory/deceased-members/20000667.html
13. Yost DM, Russell H (1944) Systematic inorganic chemistry of the fifth-and-sixth-group nonmetallic elements. Prentice Hall, New York
14. Yost DM, Russell H, Garner CS (1947) The rare-earth elements and their compounds. Wiley, New York
15. Hildebrand J (1946) Letter to DM Yost, 2/7/46, from the Yost file, #6.12, Caltech archives
16. Yost DM, Kaye AL (1933) An attempt to prepare a chloride or fluoride of xenon. J Am Chem Soc 55:3890–3892
17. Hoppe R, Dahne W, Mattauch H, Rodder K (1962) Fluorination of xenon. Angew Chem Int Ed 1:599
18. Hargittai I (2011) Drive and curiosity: what fuels the passion for science. Prometheus Books, Amherst NY, p 231–232
19. Hargittai I (2003) Candid science III: more conversations with famous chemists. Imperial College Press, London, p 33–34
20. Hyman HH, Ed. (1963) Noble gas compounds. U of Chicago Press, Chicago, p 21–22
21. Hager T (1995) Force of nature: the life of Linus Pauling. Simon & Schuster, New York, p 212
22. Extensive correspondence between DM Yost and the National Defense Research Committee and others, from the Yost file, #9.9 and 9.1.1, Caltech archives

23. Yost DM, Oppenheimer JR (1943) Exchange of letters during January November 1943, from the Yost file, #10.1, Caltech archives
24. Serafini A (1989) Linus Pauling: a man and his science. Paragon House, New York, p 218
25. Yost D, DuBridge L (1948) Exchange of memos, 3/10/48 and 6/19/48, from the Yost file, #5.3, Caltech archives
26. Taube H (1949) Letter to Yost, 11/25/49, from the Yost file, #17.3, Caltech archives
27. Jenkins WA, Yost DM (1959) On the kinetics of the exchange of radioactive hydrogen between hypophosphorous acid and water. The mechanism of the oxidation of hypophosphorous acid. J. Inorg Nucl Chem 11:297–308
28. Yost DM, Jenkins WA (1953-1959) Exchange of letters, from the Yost file, #14.6, Caltech archives
29. Yost DM (1956) Letter to S Hakomori, 11/28/56, from the Yost file, #6.4, Caltech archives
30. Yost DM (1963) Letter to LF Audrieth, 11/21/63, from the Yost file, #1.22, Caltech archives
31. Waugh JS (2009) Sixty years of nuclear moments. Ann Rev Phys Chem 60:1–19
32. Shoolery JN (2002) James N. Shoolery, interview by Arnold Thackray and David C. Brock at Palo Alto, California, 18 January 2002 (Philadelphia: Chemical Heritage Foundation, Oral History Transcript # 0230)
33. Yost DM (1959) Letter to Rev. J Duke, 10/30/59, from the Yost file, #5.4, Caltech archives
34. Anson FC (2012), personal communication
35. Yost DM (1957–1959) Letters to various ACS officers, from the Yost file, #1.6 and 1.7, Caltech archives
36. Yost DM (1963) Letter to the NSF, 5/2/63, from the Yost file, #8.21, Caltech archives
37. Yost DM (1964) Letter to L. Helmholtz, 3/30/64, from Yost file, #6.11, Caltech archives
38. Yost DM (1946) From the Yost file, #13.2, Caltech archives
39. Roller D (1954) Letter to DM Yost, 3/16/54, from the Yost files, #16.2, Caltech archive
40. Yost DM (1959) Letter to WG Parks, 4/16/59, from the Yost file, #1.8, Caltech archives
41. Yost DM (1950) Letter to WC Fernelius, 2/17/50, from the Yost file, #5.11, Caltech archives
42. Serafini A (1989) Linus Pauling: a man and his science. Paragon House, New York, p 22
43. Serafini A (1989) Linus Pauling: a man and his science. Paragon House, New York, p 63–65
44. Serafini A (1989) Linus Pauling: a man and his science. Paragon House, New York, p 33–34
45. Yost DM (1948) Letter to HC Urey, 3/29/48, from the Yost file, #12.12, Caltech archives
46. Yost DM (1954) Letter to S Winkelman, 12/6/54, from the Yost file, #13.6, Caltech archives
47. Yost DM (1973) Letter to EL Miller, 3/8/73, from the Yost file, #7.1, Caltech archives
48. Roberts JD (2012), personal communication
49. Yost DM (1947) Memo to the CCE Division, 7/28/47, from the Yost file, #2.2.3, Caltech archives
50. Yost DM (1958) Letter to HS Johnston, 8/27/58, from the Yost file, #7.1, Caltech archive
51. Pauling L (1956) Memo to DM Yost, 9/12/56, from the Yost file, #10.6, Caltech archives
52. Caltech Chemistry and Chemical Engineering (CCE) Divisional archives
53. Yost DM (1956) Memo to L Pauling, 5/29/56, from the Yost file, #10.6, Caltech archives

Chapter 5
Agents of Respectability

> *History has vindicated his judgement, but we may well ask why he could have been so sure.*
>
> Colin Russell, *The Structure of Chemistry*

The preceding discussion has placed the onset of the transition of inorganic chemistry, from a secondary, underpopulated subfield to a fully coequal special-ization, around the middle 1950s. We know "when" and can now turn to the equally interesting, and perhaps more challenging, questions of "why" and "who." A number of authors of essays and textbooks have suggested explanations, but none of them, in my opinion, is entirely satisfying. In this chapter I will offer an alternative interpretation that seems to me much more consistent with both the lack of regard for the field before the transition, and the directions in which it rapidly evolved from that point forward.

Let us begin with Nyholm, who demonstrated considerable prescience in recognizing the renaissance of inorganic chemistry even as it was taking place. But why *did* he think it was happening: as Russell asks, how could he have been so sure? In his inaugural address/essay, Nyholm identified what he considered to be the two key drivers:

> The factors primarily responsible for the modern forward-looking spirit in inorganic chemistry are two *external* developments....the growth of the theoretical techniques of quantum mechanics to an extent permitting widespread chemical application....[and] those new optical electrical and magnetic techniques of physical measurement by which structure can be investigated....It is my essential thesis that the impact of quantum mechanics and of modern physical methods of attack are the main reasons for the renaissance of inorganic chemistry, leading to the present period of rapid growth [1]. (my italics)

Subsequent commentators have concurred: not only with Nyholm's identifi-cation of a post-war renaissance of inorganic chemistry, but also with his causal factors. Some also cite the war itself, and the research it generated into nuclear chemistry and the trans-uranium elements, as another important contributor:

> In the recent past the study of inorganic chemistry has reachieved a level of interest comparable to that exhibited in the earlier phases of the development of the subject of chemistry as a whole. This renewed interest has come to pass largely because of the parallel development of...theoretical principles and experimental techniques... [2].

J. A. Labinger, *Up from Generality*, SpringerBriefs in History of Chemistry, DOI: 10.1007/978-3-642-40120-6_5, © The Author(s) 2013

The success of new bonding theories in explaining structures and spectra of inorganic complexes led to a resurgence of interest in their formation and reactions [3].

The most obvious example of renaissance during the mid 1950s to late 1960s is…inorganic chemistry. A variety of factors contributed. One was war-related research as part of the Manhattan Project….The declassification of this information in the early 1950s gave inorganic chemistry new breadth. However, the biggest factor that returned inorganic chemistry to the forefront was the way in which experimentalists using new techniques began to collaborate with theoreticians [4].

The question 'Why was inorganic chemistry beginning to become a "respectable" area of study within chemistry?' is an interesting one. [D]uring the war many aspects of inorganic chemistry had become important in the nuclear bomb programme and there was a complete reappraisal of the subject and a recognition that there was vast potential for study. This was also at a time when many of the developments in quantum chemistry that had been recognized in the late 1930s were applied to inorganic systems and provided a theoretical basis for the subject [5].

The revival of interest in inorganic chemistry in recent years, following upon the discovery of new elements produced in nuclear fission… [6].

An important spinoff of the strategic importance of nuclear chemical research during and after World War II was the enormous amount of research in inorganic chemistry that was only indirectly linked to nuclear fission. Many of the major contributors to the development of inorganic chemistry in the second half of this century began their careers in research related to nuclear chemistry and physics [7].

Note that some of the above comments—particularly the last—do not actually claim that nuclear chemistry played an important ongoing part in the renaissance. Their argument appears to be, rather, that many people were perforce brought into inorganic chemistry by wartime research, and some of those stayed in the field. Indeed as has already been noted, Geoffrey Wilkinson—by anyone's account one of the "major contributors to the development of inorganic chemistry in the second half of this century" [7]—was hired at Harvard as a *nuclear* chemist (although he quickly moved on from there, as we shall see shortly).

We can exclude any major role for nuclear chemistry beyond that aspect, though, on two grounds. First, as we have already seen, the subordinate status of inorganic chemistry—at least in US academia—had much to do with its image as a purely descriptive science, with no solid intellectual underpinnings. Adding a new set of elements did little or nothing to change that: a paucity of elements whose chemistry could be studied was decidedly *not* one of the shortcomings of pre-war inorganic chemistry! More importantly, nuclear chemistry was by no means a predominant, or even significant, component in the research programs of those who led the post-war growth of American inorganic chemistry.

The centrality of new experimental and theoretical methodology is harder to deny. Certainly they both played essential roles in nearly all the inorganic research from this time forward; no renaissance would have been possible without them. However, I find it difficult to identify them as "game-changers" all on their own, for several reasons. First, developments in both arenas, especially the experimental one, were in many ways more evolutionary than revolutionary, extending work that had been going on for some time. The use of both X-ray crystallography and infrared spectroscopy for characterization of inorganic compounds was widespread well before this point in time, although their convenience and prevalence did undergo dramatic improvement after the war. Second, applications of these techniques were of course not restricted to inorganic chemistry; they were used across the board. Why then should they have served to augment the status of inorganic chemistry *relative* to other branches of chemistry? Finally, I find it hard to accept that the major impetus came from outside the field (as suggested by Nyholm's use of the word "external" in the above quote): inorganic chemists themselves were the movers and shakers.

Dyson has recently described two complementary approaches to historical analysis, focusing on scientific progress as driven by new ideas or by new tools, which he calls "Kuhnian" and "Galisonian" respectively [8]. I take a Kuhnian stance, and argue that the main agents in the remarkable post-war growth of inorganic chemistry were conceptual, not practical (although, admittedly, the concepts I invoke may not rise to the stature of those Dyson is talking about, such as relativity and quantum mechanics!). Two key research topics, mechanism and organometallic chemistry, attracted new and intense attention, thus bringing a much higher level of intellectual rigor into the field.

Neither of these topics was entirely new, of course. There was activity in inorganic reaction mechanisms before this period, although rather sparse and scattered; but the area really took off from this point. As one author puts it: "The study of inorganic reaction mechanisms, which dates from the 1860s, is by no means new. The majority of information has, however, been obtained since the 1950s" [9]. The beginnings of this increased focus on mechanistic study as a route to systematic understanding of a complex and confusing mass of data can be traced back a little earlier, to John Bailar's revival of coordination chemistry. A biographical essay on the occasion of Bailar's death quotes him on his early inspiration [10]:

> "In 1893 Paul Walden [1863–1957] discovered the very interesting inversion reaction which bears his name. It was an extremely important discovery, for it called attention to the chemists of that day that reactions have mechanisms. It occurred to me that if we repeated Werner's experiment….we might also get an inversion…if we could get an inversion with an octahedral model rather than a tetrahedral one, we might be able to rule out some of the theories which had been advanced for the inversion in reactions of the tetrahedral organic molecules." Near the end of his career John still regarded this work as being his most significant scientific achievement.

Bailar's students, especially Fred Basolo, did much to spread the mechanistic word throughout the US inorganic community. As noted earlier, Basolo went off to Northwestern, and there recruited physical chemist Ralph Pearson (Fig. 5.1) to join him in promulgating mechanism as the fulcrum of a research program in coordination chemistry [11]:

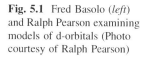

Fig. 5.1 Fred Basolo (*left*) and Ralph Pearson examining models of d-orbitals (Photo courtesy of Ralph Pearson)

Ralph [G.] Pearson, whose training was in kinetics and mechanisms of reactions, was working on organic compounds with some of our organic chemists and helping them learn this kind of chemistry. I would always say to him, "How about coming on board? No one else is looking at the systems we would be studying. We could work on what you're doing on organic compounds in the metal complex class of compounds. Almost anything we could do would be original, publishable, and useful as far as students are concerned to get their Ph.D. theses approved because that also would be original work." Pearson did come on board at Northwestern, and together, we did what was not possible individually. He needed me, I needed him, and we arrived there at just the right time.

One of Basolo and Pearson's major early successes was a correct explanation of why displacement of chloride in many coordination compounds, such as the classic Werner complex $[Co(NH_3)_5Cl]^{2+}$, is much faster under basic than acidic conditions. Rather than reflecting the fact that $[OH]^-$ is a better nucleophile than H_2O, as Nyholm and others had suggested, they demonstrated that a (previously proposed) facile "conjugate base" mechanism (Fig. 5.2) operates. In the late 1950s they codified a substantial body of such work, by themselves and others, in one of the first comprehensive treatises on inorganic mechanisms [12]. Their book played a significant part in making the chemistry community aware of the accomplishments and (especially) the opportunities in the field.

Two others who played key roles in making mechanistic study central to inorganic chemistry were Henry Taube and Jack Halpern (Fig. 5.3). Taube (1915–2005) began his academic career at Cornell in the 1940s, moving from there to Chicago and subsequently to Stanford. He was a pioneer in a number of areas, including the mechanism of electron-transfer reactions and the use of isotopic tracers for study of those and other inorganic transformations. He devised the first unequivocal demonstration of the so-called inner-sphere mechanism of electron

$$[Co(NH_3)_5Cl]^{2+} + OH^- \rightleftharpoons [Co(NH_3)_4(NH_2)Cl]^+ + H_2O$$

$$[Co(NH_3)_4(NH_2)Cl]^+ \longrightarrow [Co(NH_3)_4(NH_2)]^{2+} + Cl^-$$

$$[Co(NH_3)_4(NH_2)]^{2+} + H_2O \longrightarrow [Co(NH_3)_5(OH)]^{2+}$$

$$[Co(NH_3)_5Cl]^{2+} + OH^- \longrightarrow [Co(NH_3)_5(OH)]^{2+} + Cl^-$$

Fig. 5.2 The S_N1CB mechanism for rapid base hydrolysis of cobalt ammine chloride complexes

Fig. 5.3 Pioneers of mechanistic inorganic chemistry: Henry Taube (*left*) and Jack Halpern (*right*), ca. 1960s (Photos courtesy of the Department of Chemistry, University of Chicago)

transfer, which proceeds via an intermediate wherein a ligand on one of the participants of the redox reaction acts as a bridge between the two metal centers [13]. The example of Fig. 5.4 shows a reaction between a trivalent cobalt complex containing a chloride ligand and a divalent chromium complex; the chloride winds up in the coordination sphere of the trivalent chromium product. Carrying out the reaction in the presence of an excess of radiolabeled chloride leads to no incorporation of label in the product, proving that the chloride must have gone directly from one metal to the other rather than being liberated into solution and then finding its way to the product. For this and a large body of similarly seminal work, Taube received the 1983 Nobel Prize in chemistry [14].

Halpern (1925–), like Taube, started as a physical chemist. He had no exposure to inorganic chemistry whatsoever as an undergraduate or graduate student, and his first academic position was in the department of metallurgy (!) at British Columbia. But he turned his long-standing interest in mechanism to problems involving metal complexes, an important focus of his long career that continued after he moved to Chicago as Taube's replacement [15]. He took particular interest

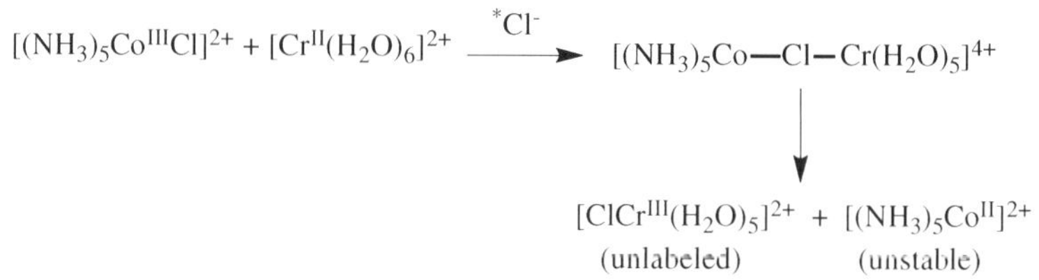

[(NH3)5CoIIICl]2+ + [CrII(H2O)6]2+
*Cl-
[(NH3)5Co—Cl—Cr(H2O)5]4+
[ClCrIII(H2O)5]2+ + [(NH3)5CoII]2+
(unlabeled)
(unstable)

Fig. 5.4 A typical experiment used by Taube to establish the existence of an inner-sphere mechanism for redox reactions of transition metal complexes

in the activation of molecular hydrogen, a subject which by virtue of its applicability to catalytic hydrogenation was important in linking the two main themes of mechanism and organometallic chemistry. Probably his most memorable contribution was the elucidation of the mechanism of asymmetric hydrogenation, whereby H_2 is added to a prochiral olefin in the presence of a chiral catalyst to give a single enantiomer of the saturated product. It had generally been assumed that selectivity in such processes results from an enzyme-like "lock-and-key" situation, in which steric interactions force the catalyst to bind preferentially to one face of the olefin, and the major product is derived from that intermediate. Halpern showed that while there was indeed a strong binding preference, the major product actually resulted from the *less*-preferred isomer of the intermediate, because it reacted *much* faster with H_2 to complete the hydrogenation (Fig. 5.5) [16]. Such analyses have helped asymmetric catalysis become one of the dominant tools of modern synthetic organic chemistry.

Turning now to organometallic chemistry: it was not new either, originating primarily with the zinc alkyls first discovered by Frankland around 1850 (see Chap. 2). Over the next century that chemistry was extended to a wide range of *main-group* metals, most prominently the alkylmagnesium reagents (RMgX) developed by the eponymous Victor Grignard in the early twentieth century. But there were hardly any analogous compounds of *transition metals*; attempts to synthesize them invariably led to decomposition. In 1951 Pauson attempted to exploit that behavior, anticipating that a bis(cyclopentadienyl)iron compound would be highly unstable, liberating a convenient precursor to fulvalene, his ultimate target (Fig. 5.6). Instead, he obtained an extremely stable orange compound, which he suggested *was* the organoiron compound $(C_5H_5)_2Fe$, although no explanation for why it should be so stable was readily apparent.

Keq
major intermediate
H2
k
minor product
catalyst + olefin
K'eq
minor intermediate
H2
k'
major product

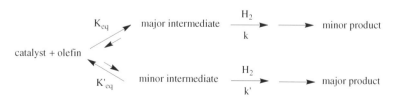

Fig. 5.5 Schematic mechanism for asymmetric catalytic hydrogenation. The stereochemistry of the preferred intermediate ($K_{eq} > K'_{eq}$) does not govern that of the preferred product ($k' \gg k$)

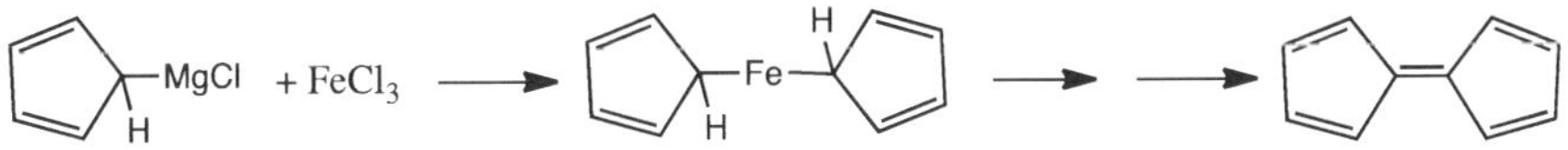

Fig. 5.6 Pauson's planned route to fulvalene, showing the structure originally proposed for the unexpectedly stable organoiron complex that was obtained instead

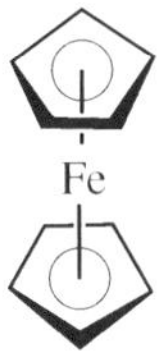

Fig. 5.7 Actual structure of ferrocene, the product obtained from the reaction shown in Fig. 5.6

The correct interpretation was given in the following year, by Geoffrey Wilkinson and R. B. Woodward at Harvard. Instead of a simple structure with two carbon-iron bonds, they recognized that a "sandwich" structure (Fig. 5.7), utilizing all the π-electrons of the cyclopentadienyl rings, could account for the great stability (as well as the subsequently discovered chemistry) of the compound [17]. Ferrocene, as the compound was named, ignited interest in organotransition metal chemistry, led by Wilkinson in the US[1] and E.O. Fischer at Munich; the two shared the 1973 Nobel Prize for their work in the field.

The organometallic chemistry of the transition metals[2] quickly became one of the dominant areas of inorganic chemistry; Helmut Werner (no relation to Alfred) has written a detailed, though somewhat selective and personal, history of the field [18]. Several developments over the next decade or so were particularly significant in consolidating its role as an agent of respectability. James Collman (Fig. 5.8) began to codify the initially diverse and confusing array of organometallic reactions. Influenced by Taube, whom he got to know well while on sabbatical at Stanford in 1965 [19], Collman brought mechanism-based organization to the field, identifying ubiquitous patterns of reactivity such as oxidative addition, insertion, etc. That contribution can be readily traced in the evolution of organometallic textbooks. Many of the earlier popular texts were arranged according to the Periodic Table [20] or ligand type [21], organizational schemes that are far

---

[1] Wilkinson moved back to the UK in 1956; Gordon Stone continued to lead a program in organometallic chemistry at Harvard, but also left for the UK. Both moves were largely a consequence of the general disdain for inorganic chemistry at Harvard, as documented in Chap. 4.

[2] Main-group organometallic chemistry continued to reside primarily in the realm of organic chemists, as it had since Frankland. It is not entirely obvious why that should be so, or why inorganic chemists had apparently been unable to stake a claim to any share of Frankland's discoveries a century earlier. From the 1950s on, though, boundary lines between inorganic and organic territories of organometallic chemistry were gradually dissolved, particularly by widespread applications of organotransition metal complexes as reagents and catalysts for organic synthesis.

Fig. 5.8 Jim Collman
(1932–), captured giving a
lecture on organometallic
chemistry, ca. 1972 (Photo
courtesy of Jim Collman)

better suited to description than to explanation. In contrast, the important 1980 text
by Collman and co-author Lou Hegedus focused primarily on those mechanistic
classifications, providing a much greater degree of interpretive and predictive
power [22].

Around the same time that Collman's systematization of the field began to gain
attention, a number of chemists who specialized in mechanistic study turned their
expertise to problems in organometallic chemistry. Halpern in particular extended
his program on hydrogen activation to organometallic complexes and catalytic
hydrogenation, achieving remarkably detailed understanding of highly compli-
cated systems, as shown above in Fig. 5.5. Also, several physical organic chemists,
many with no formal training in inorganic chemistry,[3] made central contributions
by applying their armament of concepts and techniques to this new realm.

To repeat, Nyholm and others were clearly not wrong in pointing to new
experimental and theoretical tools as playing a key role in the explosive growth of
inorganic chemistry. In my view, though, that should be seen as an *auxiliary*
role—necessary but not sufficient—that served mainly to facilitate work in the
fields that finally made inorganic chemistry truly respectable. Halpern, one of the
leaders of the mechanistic thrust, suggests that the developments in theory and

---

[3] Among the most prominent are Bob Bergman (who began his career at Caltech before moving
to UC Berkeley), Chuck Casey (Wisconsin), and Maurice Brookhart (North Carolina). Bob
Grubbs, who won the 2005 Nobel Prize in Chemistry for his work on olefin metathesis (shared
with Dick Schrock, who like me was an academic grandson of Wilkinson via John Osborn) *did* do
an inorganic postdoc, with Collman, before starting his independent career at Michigan State and
then moving on to Caltech.

even in nuclear chemistry should be considered mainly as *enabling* technologies; he notes that the emergence of ligand field theory provided a framework for explaining much of the mechanistic findings in substitution reactions of coordination complexes, while the availability of radioisotopes generated powerful new methods for following electron-transfer reactions [15]. But the turn to mechanism that Bailar and his school initiated at least partially predates both of those developments. Similarly, the rapid progress in organometallic chemistry following the report of ferrocene would have been inconceivable without NMR; but NMR was just beginning to be routinely accessible to academic researchers in the first half of the 1950s, when Wilkinson and Fischer began their pioneering work.

With its newfound emphasis on mechanism, as well its incorporation of the boundary field of organometallic chemistry, inorganic chemistry rapidly became a much more intellectually attractive field within academia. Early on there was the inauguration of a Gordon Research Conference (GRC) on inorganic chemistry, a series which began in 1951, due mainly to the efforts of Bailar and some of his students [11]. Mechanism and organometallic chemistry quickly became major discussion topics; by the third meeting, in 1953, they accounted for 75 % of the organized sessions, and one or both were designated topics in most of the subsequent meetings. Eventually organometallic chemistry became so prominent that a new GRC on that topic was established in 1972.[4]

Next was the establishment of a separate Division of Inorganic Chemistry of the ACS; previously there had been only a joint physical-inorganic division. Again Bailar was the main mover. Starting in 1956 he organized a grass-roots movement of inorganic chemists who were readily convinced that the time for independence had finally come, and the new division was established, with Bailar as its first chairman, in 1958. Membership grew rapidly: as we saw in Chap. 3, the inorganic division has accounted for the largest representation (in terms of numbers of presentations) at most ACS national meetings from the late 1970s on [23].

The specialist ACS journal for the field, *Inorganic Chemistry*, appeared around the same time, although that development had a somewhat more tortuous history. In the early 1950s, recognizing a need for an English-language complement to the *Zeitschrift für anorganische und allgemeine Chemie*, the ACS considered but decided against founding such a journal. Instead Pergamon Press in the UK founded a new *Journal of Inorganic and Nuclear Chemistry* (*JINC*) [24]. A few years later, when the strong growth of the field led the ACS to revive the idea, there was considerable resistance on the grounds that it would inevitably undermine the position of *JINC*. I. R. Maxwell, the Managing Director of Pergamon, wrote: "*JINC* was originally started at the suggestion and with the assistance of leading American inorganic and nuclear chemists and colleagues in England and all over the world warmly welcomed this American idea. It would, therefore, not

---

[4] A separate GRC on Inorganic Reaction Mechanisms split out much later, in 1997; that delay probably results from the fact that mechanism remained such an intimate component of the mainstream Inorganic Chemistry GRC.

be understood why the administrative officials of the American Chemical Society should wish to start up a competing journal" [25]. Indeed, as the latter implies, the move seems to have been—at least at first—primarily an initiative of ACS administrators, *not* American inorganic chemists. The Secretary-Treasurer of the inorganic division told Maxwell: "I am afraid that there is quite a bit of truth in the rumor that the American Chemical Society is thinking of starting a Journal of Inorganic Chemistry….it is no longer a question of should one be started but rather when should it be started. The Executive Committee of this Division has consistently stated that such a Journal is not needed in view of present arrangements, but to no avail" [26]. However, the Executive Committee soon came around [27], and the US inorganic community enthusiastically supported the new venture. The first issue appeared in 1962, under the editorship of Robert Parry (1917–2006), one of Bailar's most prominent students, and *Inorganic Chemistry* has been a resounding success ever since [28].[5]

I have proposed that the dual focus on mechanism and organometallic chemistry, starting around 1950, was the major causative factor of the ensuing dramatic progress of inorganic chemistry, as documented in Chap. 3. But why should that have been so? and perhaps more importantly, can we show that it *was* so? First why: we have seen in earlier chapters the pre-renaissance view of inorganic chemistry: intellectually inferior to the more-respected fields of organic and physical; typified by phenomenological descriptions as opposed to explanatory science; perhaps not even a "real" specialization, not clearly differentiated from general chemistry. The mechanistic turn provided a clear escape route from that lowly status. Inorganic chemists could now be seen to be seeking the same level of detailed fundamental understanding as their organic and physical colleagues.

Simultaneously, the growth of organometallic chemistry played a major role in claiming a share of the traditionally much more positive image of organic chemists, especially after organometallic chemists turned strongly to mechanism as well. The potential advantages of that move were recognized very early on. In 1958 a committee of the then-newly formed ACS Division of Inorganic Chemistry proposed formation of a subdivision for organometallic chemistry *within* itself, arguing that "the Inorganic Division would be strengthened and enlarged by bringing in all those from other fields who are interested in organometallics, a subject that crosses many traditional boundaries" [29]. (Such a subdivision *was* eventually established, but not until nearly 10 years later.) It is noteworthy that the afore-mentioned Gordon Research Conference in Organometallic Chemistry was primarily run by (and for?) inorganic chemists, at least initially. Of the 38 major talks at the first three conferences (1972, 1974 and 1976) 28 were given by inorganic chemists, and another 7 by people who started their careers as organic chemists but switched to calling themselves inorganic chemists (or both) at about this time. The majority of the latter, as well as *all* the talks given by "pure"

---

[5] It *did* however have the anticipated negative effects on *JINC*, which has since become a considerably less prestigious journal.

Fig. 5.9 Four generations of the Bailar descent, photographed in 1983 on the occasion of the 3rd Oesper Symposium at University of Cincinnati honoring Fred Basolo. Founding father John Bailar (*4th from left*); his student Fred Basolo (*4th from right*); Basolo's long-time collaborator Ralph Pearson (*1st on left*); their students Harry Gray (*2nd from left*), Andrew Wojcicki (*3rd from left*) and Robert Angelici (*2nd from right*); and Gray's student Mark Wrighton (*3rd from right*); along with a representative of the Oesper family (*1st on right*) (Photo courtesy of the Oesper Collections, University of Cincinnati)

organic chemists, dealt with main-group organometallic chemistry, whereas all the inorganic chemists spoke on organotransition metal chemistry, reflecting the effective division of labor remarked upon above.

The clearest demonstration that mechanism and organometallic chemistry *were* the major factors can be found in the academic "ancestry" of the US inorganic faculty following the period of rapid growth. I have identified, above, a small number of pioneers in each of the two areas in the US. For mechanism: Bailar and his students, especially Basolo (the most numerous branch of this genealogy; see Fig. 5.9 for some prominent members), Taube and Halpern; for organometallic chemistry: Wilkinson, Stone and Collman. The "family tree" of Fig. 5.10 shows the numbers (and names of some of the more prolific mentors) of their academic descendants—graduate students and postdocs—who held faculty positions in US PhD-granting departments in 1983;[6] they add up to 285, out of a total of 566 who are identified as inorganic chemists. Thus over 50 % of the total can be traced to those six lines of descent for their graduate or postdoctoral training (or both, in more than a few cases!). The numbers are even more striking if we focus on the most prestigious of those institutions: a "top ten" ranking of chemistry

---

[6] The year was chosen somewhat but not entirely arbitrarily; it represents a point where a large majority of faculty would have gotten their training during or after the 1950s, but before the number of successive intermediate generations would have made tracing lines of descent excessively onerous. Also see Notes on quantitative methodology.

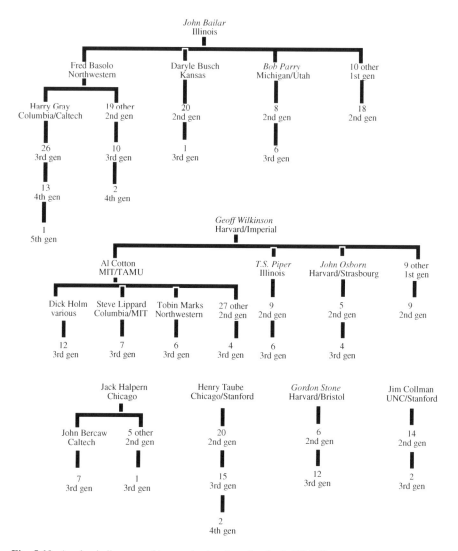

Fig. 5.10 Academic lineages of inorganic chemistry faculty in US PhD-granting departments, as listed in the 1983 DGR. Names in italics were no longer in one of those departments, having by then retired (Bailar, Parry), died (Piper), or taken a position abroad (Wilkinson, Stone, Osborn)

departments [30] was comprised of Harvard, University of California Berkeley, Caltech, MIT, Columbia, Stanford, Illinois, UCLA, Chicago and Cornell. In the 1983 DGR those departments listed 47 inorganic chemists, of whom 38, around 80 %, are represented in those six lineages.

The same dominance can be seen in surveys of award winners. The ACS Award in Inorganic Chemistry and the ACS Award for Distinguished Service in the Advancement of Inorganic Chemistry have been presented annually since 1962 and 1965 respectively. There have been 80 awardees in all,[7] but a number of them did their entire academic training out of the US or before these lines were established.[8] Those awardees who received their Ph.D. in 1955 or later, and did their graduate study and/or postdoctoral work in the US, total 44; the six lineages account for 34 of these, again around 80 %. The statistics for the ACS Award in Pure Chemistry are still more dramatic: of the 17 inorganic chemists who have received this award since 1970 (see Fig. 3.4), *all but one* descend from these six founding programs.

These statistics establish beyond doubt that this cohort of chemists, whose academic descent can be traced to the groups that led the moves into mechanistic and organometallic studies, came to dominate the resurgent field of inorganic chemistry, both quantitatively and qualitatively. Such a finding should go a long way towards confirming the hypotheses that the renaissance was primarily driven by concepts, not tools;[9] and that the crucial concepts have been correctly identified.

Of course, those two topics by no means constituted the totality or even the majority of inorganic research programs; more traditional areas of study remained strong, and new ones began to appear and burgeon within the field. How then did inorganic chemistry maintain its new-found status, not to mention its coherence, in the face of potentially fragmenting forces? That question will be the topic of the next, concluding chapter.

References

1. Nyholm RS (1957) The renaissance of inorganic chemistry. J Chem Educ 34:166–169
2. Lagowski JJ (1973) Modern inorganic chemistry. Marcel Dekker, New York, p 111
3. Cobb C, Goldwhite H (1995) Creations of fire: chemistry's lively history from alchemy to the atomic age. Plenum Press, New York, p 372
4. Kieffer WF (1980) Sputnik, trickle-down and renaissance. J Chem Educ 57:31–32
5. Lord Lewis of Newnham, Johnson BFG (1997) Cyril Clifford Addison: 28 November 1913-1 April 1994. Biographical Memoirs of Fellows of the Royal Society 43:2–12
6. Wood CW, Holliday AK (1960) Inorganic chemistry: an intermediate text. Butterworths, London, p vii
7. Butler IS, Harrod JF (1989) Inorganic chemistry: principles and applications. Benjamin/ Cummings, Redwood City, CA, p 5
8. Dyson FJ (2012) Is science mostly driven by ideas or by tools? Science 338:1426–1427.
9. Cooke DO (1979) Inorganic reaction mechanisms. The Chemical Society, London, p iii

---

[7] There have actually been 100 awards from 1962–2013, but only 80 distinct recipients, since 20 people have earned both.

[8] The Distinguished Service Award (affectionately known by some as the "Old Farts' Award") is, not surprisingly, particularly represented by older chemists.

[9] I have not carried out a similar analysis based on descent from groups noted for the introduction of new experimental and theoretical tools, but it seems certain that no remotely comparable concentration could exist.

10. Kauffman GB, Girolami GS, Busch DH (1993) John C. Bailar Jr. (1904–1991): father of coordination chemistry in the United States. Coord Chem Rev 128:1–48
11. Basolo F (2002) Fred Basolo interview by Arnold Thackray and Arthur Daemmrich at Northwestern University, Evanston, Illinois, 27 September 2002 (Philadelphia: Chemical Heritage Foundation, Oral History Transcript # 0264)
12. Basolo F, Pearson RG (1958) Mechanisms of inorganic reactions: a study of metal complexes in solution. John Wiley and Sons, New York
13. Taube H, Myers H, Rich RL (1953) J Am Chem Soc 75:4118–4119
14. van Houten J (2002) A century of chemical dynamics traced through the Nobel Prizes: 1983: Henry Taube. J Chem Ed 79: 778–790
15. Halpern J (2012) Personal communication
16. Halpern J (1982) Mechanism and stereoselectivity of asymmetric hydrogenation. Science 217:401–407
17. Laszlo P, Hoffmann R (2000). Ferrocene: ironclad history or Rashomon tale? Angew Chem Int Ed 39:123–124
18. Werner H (2009) Landmarks in organo-transition metal chemistry: a personal view. Springer, New York
19. Collman JP (2012) Personal communication
20. King RB (1969) Transition-metal organometallic chemistry: an introduction. Academic Press, New York
21. Green MLH (1968) Organometallic compounds volume two: the transition elements. Methuen & Co., London.
22. Collman JP, Hegedus LS (1980) Principles and applications of organotransition metal chemistry. University Science Books, Mill Valley, CA
23. Bailar JC Jr. (1989) A history of the Division of Inorganic Chemistry, American Chemical Society. J Chem Ed 66:537–545.
24. Katz JJ (1960) Letter to L. P. Hammett, 4/1/60, from the Yost file, #1.8, Caltech archives
25. Maxwell IR (1960) Letter to J. J. Katz, 5/18/60, from the Yost file, #1.8, Caltech archives
26. Christensen EL (1960) Letter to I. R. Maxwell, 4/4/60, from the Yost file, #1.8, Caltech archives
27. Brimm EO (1960) Letter to I. R. Maxwell, 7/18/60, from the Yost file, #1.8, Caltech archives
28. Eisenberg R (2011) *Inorganic Chemistry*, volume 50: a golden year. Inorg Chem 50:1–3
29. Niche for organometallics. Chem Eng News 36, 7 April 1958: 105
30. Gorman J (1993) The Gourman report: a rating of graduate and professional programs in American and international universities: leading graduate chemistry programs, 6th ed. http://consusrankings.com/index.php?s=chemistry. Accessed 6 March 2013

Chapter 6
Conclusions

> *Things fall apart; the centre cannot hold;*
> *Mere anarchy is loosed upon the world*
> W. B. Yeats, *The Second Coming*

In the preceding chapters I have documented a dramatic upsurge in both the numbers and reputations of inorganic chemists in US academia during the years following what Nyholm identified (correctly, I think we would all agree) as the renaissance of the 1950s. It is also clear that the lion's share of that cohort descends from a small number of pioneering groups in the two areas that I argue were central to that renaissance: mechanism and organometallic chemistry. Of course, those were by no means the only research topics in the field, or necessarily the most popular at any given point in time. More "traditional" (pre-renaissance) activities continued unabated, and inorganic chemistry diversified even further, spreading across old boundaries to become ever more interdisciplinary. One might have thought that such developments would constitute a centrifugal force, tending to favor fragmentation. On the contrary, as the numbers show, inorganic chemistry has retained and even enhanced its coherent, intellectually viable persona right up to the present. In this concluding chapter I briefly discuss the evolution of the field in the latter part of the twentieth century within the general context of discipline formation.

Study of the formation of scientific disciplines is a topic much too broad for thorough consideration here. There is a large body of literature from a wide range of viewpoints, of which I cite here only a few examples: collections of essays that variously take historical/philosophical [1, 2], sociological [3], or cultural [4] approaches. As a basis for comparison we might take a brief look at physical chemistry at the end of the nineteenth century, a case that has received considerable attention both as a particular historical episode and as an exemplar of discipline formation [5–10].

Mary Jo Nye contrasts the emergence of specialization in organic and physical chemistry. On the one hand "[D]uring the course of the nineteenth century, chemists doing organic chemistry took over the discipline of chemistry" whereas "By the 1880s and 1890s, some scientists began to define their work in chemistry as 'physical chemistry' *against* organic chemistry"—descriptions in language that evokes conquest and secession respectively [11]. The latter move was unquestionably successful —three of the first four Nobel Prizes in chemistry were awarded to chemists associated with the then-new subfield—and quickly spread to

J. A. Labinger, *Up from Generality*, SpringerBriefs in History of Chemistry, DOI: 10.1007/978-3-642-40120-6_6, © The Author(s) 2013

the US. "[B]y the mid-1920s, physical chemists like [G. N.] Lewis were beginning to challenge organic chemists for leadership in the American chemical community" [12].

Should we talk of inorganic chemistry around the same time period as an identifiable, separate discipline? Despite its conflation with general chemistry, as discussed in Chap. 1, I think the answer must be yes; but we would need to describe its separation in terms more like "surrender" or "retreat." It was a negative move, not a positive one. As one indication of that claim, Nye (and others as well) consider the identification of a founding group and genealogy as a crucial element in a paradigm of discipline formation. For physical chemistry the central figures would include van't Hoff, Ostwald, Arrhenius, and a few others [6]. But there are no obvious candidates for any analogous founding group of inorganic chemists up to 1900, or even up to 1950. Other important markers, such as establishment of institutes and chairs, professional organizations, journals, international meetings, etc., are well represented in late-nineteenth century physical chemistry [7]; but again, these are largely absent (with perhaps the sole exception of the founding of *ZAC*) from inorganic chemistry of that time. Perhaps, to use a distinction Lenoir offers [4], it would be more appropriate to talk about the early history of inorganic chemistry as a research program, but *not* a disciplinary program.

In contrast, the history of inorganic chemistry from the 1950s onwards *does* look a lot like classic discipline formation, despite the fact that no truly *new* discipline was being formed. In an essay [13] aimed at constructing a general theory about the formation of scientific disciplines (which they call "Scientific/Intellectual Movements," or SIMs) and comparing them to social movements in general, two sociologists offer several criteria for identifying a SIM, along with "four general propositions for explaining the social conditions under which SIMs are most likely to emerge, gain prestige, and achieve some level of institutional stability." While inorganic chemistry does not satisfy all of their defining criteria (which include that a SIM must substantially break with past practice and have a finite lifetime), their four general propositions (I paraphrase them below) do seem strikingly applicable to the developments we have been examining:

Proposition 1 There must be significant dissatisfaction with the current state of affairs.

We have seen that a relatively small group of inorganic chemists, led by Bailar and his students, were indeed aggrieved: not about what inorganic chemists were doing—there were no real "dramatic breaks with past practice"—but by the inadequate recognition, from both within and without, of the merits and independent status of their field. That dissatisfaction inspired them to instigate strong and effective action to overcome it, particularly the establishment of the new ACS Division of Inorganic Chemistry, and, a little later, the journal *Inorganic Chemistry*.

Proposition 2 Structural conditions should exist that make key resources readily available.

Proposition 3 Access to micromobilization contexts plays an important role.

These are closely connected and most conveniently discussed together. Among key resources, Frickel and Gross list opportunities for employment and publication, organizational resources, and, of course, financial support. "Micromobilization contexts" is a term from the social movements literature, referring to "local sites in which representatives of the movement and potential recruits can come into sustained contact with one another." These include conferences, research retreats, and academic departments. Clearly there is much overlap between these two concepts; in particular, academic departments may (do, in the case of inorganic chemistry) play the most important role in generating employment and organizational support as well as functioning as the main sites where interactions and recruitment take place.

It is surely no coincidence that inorganic chemistry in the US took off during the post-Sputnik era, a time when university faculties were expanding, as was federal funding for research. Both of those conditions fostered an environment that strongly favored growth: those who already had academic positions were able to support ever-larger groups of students and postdocs, while the latter could feel relatively comfortable about taking a chance on a new and emerging field, knowing that there would likely be academic openings for them as well.

Proposition 4 The SIM must construct an intellectual identity for itself.

This is an argument I have already offered: that inorganic chemistry made a significant turn starting in the 1950s—from mostly descriptive and phenomenological to much more mechanistic and explanatory—that substantially enhanced its intellectual appeal. I suggest that the field's ability to sustain its identity and vitality in the face of potentially fragmenting forces—what one commentator calls "the ongoing practice of appropriation across disciplinary boundaries, always in tension with the demands of intradisciplinary coherence" [10]—may be best understood in the context of this last point.

First let us briefly survey the expansion of post-1950s inorganic chemistry. After mechanism and organotransition metal chemistry added to what was already quite a diverse range of topics that comprised inorganic chemistry, the field became ever broader from that point on. The most important new focus was on bioinorganic chemistry—primarily, the role of metals in biological systems. Jim Collman was one of the leaders of this biological turn, noted particularly for his work on synthetic models for oxygen binding and activation by proteins; he attributes much of his newly-awakened interest in the field to Taube's work on electron-transfer mechanisms [14]. Indeed, most of those who led this new move were trained in other areas of inorganic chemistry, rather than coming from a biochemical background. Harry Gray and Ken Raymond came out of the Basolo/Pearson mechanistic tradition; two other key figures, Dick Holm and Steve Lippard, are from the Wilkinson/Cotton line of descent (see Fig. 5.10).

Just as in the renaissance of inorganic chemistry in general, personal experiences appear to have played important roles in motivating some of these career redirections. Holm recalls that he got into bioinorganic chemistry after going to a conference on the subject at the instigation of a biochemist colleague, just at the time he was looking about for something new to work on [15]. Joan Valentine[1] tells a similar story: while working in the area of metal-dioxygen complexes in the 1970s, a field she felt was becoming somewhat overpopulated, a colleague told her of having heard an exciting talk on superoxide dismutase, a metalloenzyme highly relevant to dioxygen metabolism; that experience helped convince her to move into bioinorganic chemistry [16]. Gray was first attracted to bio-relevant chemistry when, while on the faculty at Columbia, he was asked to supervise a graduate student at nearby Rockefeller University, an almost entirely bio-oriented institution [17]. Lippard, on the other hand, knew he wanted to work at the interface of chemistry and biology even before he started his graduate work in the early 1960s. However, since there were few faculty working in bioinorganic chemistry in US academia at the time—none at the particular institutions he was interested in—he chose a non-biological project for his doctoral thesis, and took up bioinorganic as soon as he began his independent career [18]. None of this is to say, of course, that the development of bioinorganic chemistry would have been significantly different without these more-or-less accidental experiences; but they remind us, as we saw in Chap. 4, that individual contingencies need to be taken into account along with general trends.

Within the subfield of inorganic chemistry, several more such "sub-subfields" rose to prominence—particularly ones related to what was initially most commonly thought of as "solid state chemistry," then more popularly "materials science;" today, a good deal of that work has come under the large "nanoscience" tent. Their growth may be readily seen by the establishment of subdivisions within the ACS Division of Inorganic Chemistry, as well as of specialized journals and conferences. Subdivisions for organometallic, solid state, and bioinorganic chemistry, nanoscience, and coordination chemistry (a rather "retro" one) were initiated in 1967, 1972, 1985, 2003 and 2012 respectively; ACS journals of organometallic and materials chemistry in 1982 and 1989 respectively.[2] And each of these topics spawned further interdisciplinary diffusion by dint of their applicability: organometallic chemistry to catalysis, organic synthesis, polymer science; bioinorganic chemistry to medicine; materials science to catalysis, separation technology, devices; and many more.

---

[1] Joan Selverstone Valentine (1945-) began her career as an inorganic chemist at Rutgers, but moved to UCLA (in 1980) as a member of the Division of Biochemistry. Her work has centered on the biochemistry of metals and reactive oxygen species, with particular focus on superoxide dismutase and its suspected role in some forms of ALS (also known as Lou Gehrig's disease).

[2] While there is no ACS journal specifically devoted to bioinorganic chemistry, the phrase "including bioinorganic chemistry" appears on the cover and masthead of the journal *Inorganic Chemistry*; no other subfield is explicitly mentioned in that fashion. There are several dedicated bioinorganic journals published by other organizations, dating back to the 1970s. The journal *ACS Nano* was established in 2007; its range extends beyond the boundaries (to the extent that there are any!) of inorganic chemistry.

The evolving place of inorganic chemistry in the history of the Gordon Research Conferences is particularly interesting. The "parent" Inorganic Chemistry GRC began in 1951, and examination of nearly any one of the annual programs since then demonstrates the diversity of the field, ranging from mainstream topics (coordination compounds, lanthanide and actinide chemistry, photochemistry, electrochemistry, etc.) to more exotic subjects such as cosmochemistry and the origin of life. Some of the more popular session titles include physical methods of characterization, organometallic chemistry, materials and solid-state chemistry, metal clusters, and (especially) bioinorganic chemistry. More specialized GRCs soon appeared: Organometallic Chemistry in 1972; Metals in Biology in 1978;[3] Solid State Chemistry in 1980; Inorganic Reaction Mechanisms in 1997.

Most strikingly of all, in 2010 a tentative decision was made to phase out the Inorganic Chemistry GRC, which had taken place every year (save one) since 1951, on the grounds that the field had become *too* diverse. Dwindling attendance had led the GRC administration to conclude that putting together a program that could keep the interest of most of the attendees for most of the conference might no longer be practicable. This is precisely the sort of development that my comments on "fragmenting forces" might have led one to anticipate. However, the inorganic community strongly resisted the move: following a widespread appeal, the conference was reinstated (now on a biennial schedule) beginning in 2012, and attendance recovered [19]. Clearly something strongly holds the community of inorganic chemists together, even as their specific research topics move farther and farther apart.

What does hold a community together? Law, in what might be considered as an alternate formulation of Frickel and Gross's 4th general principle (see above), contrasts groups of scientists linked by mutual interest in certain research problems—what he terms a relationship of "organic solidarity"—to those jointly committed to certain models of explanation—a relationship of "mechanical solidarity." He suggests that "the development of a specialty may be seen as constituting a move from an organic to a mechanical basis of solidarity" [20]. That conceptual framework can perhaps help us understand *both* the renewed interest in inorganic chemistry beginning in the 1950s *and* the growth and stability of the field throughout the remainder of the twentieth century (and the twenty-first, so far). The "mechanical move" of substituting mechanism and an explanatory mode of research for phenomenological study of certain classes of problems as a key to unifying the field enabled inorganic chemistry to gain the intellectual respect needed to reach parity with organic and physical chemistry, as well as engendering a "centripetal" force, capable of maintaining that unity even as the scope of topics continued to balloon.

---

[3] This conference had existed under the name of Metals and Metal Binding in Biology since 1962, but those earlier conferences had much less chemistry content. There have been later GRCs on "sub–sub-subfields" within bioinorganic chemistry: Mo- and W-containing enzymes (1999–2009); vitamin B12 (2003–2009); iron-sulfur enzymes (1994-ongoing).

One locus where we might trace that move is inorganic textbooks (a sketch of such an approach, applied to materials science, has appeared elsewhere [21]). Attempting to do a comprehensive study along such lines would greatly exceed the scope of this book, but we can look at a couple of examples. In a typical early (1920s) text [22] (written however by a self-described Professor of Organic Chemistry!), over half the chapters consist of descriptive chemistry arranged by elemental groupings, while most of the remaining chapters would be equally (or more) appropriate in a general chemistry textbook. By the 1950s, changes in the preferred style of presentation are readily apparent. Therald Moeller's 1952 text, which Bailar (Moeller's colleague at Illinois) called "the first book that was really designed as a modern text" [23], is roughly evenly split (in chapters, if not in page length) between "Principles" and "The Chemical Elements" with the latter, as usual, organized by periodic group [24]. Still later books emphasize principles to a much greater extent, such as the 1994 text by Shriver et al. [25], which devotes less than a third of its length to an exposition of descriptive chemistry.

To be sure, not every inorganic chemist views this trend positively: some dissatisfaction with the perceived neglect of descriptive chemistry may be detected in post-renaissance texts and elsewhere. Indeed, one 1960 text [26] is *entirely* descriptive chemistry, arranged by periodic group. Such rethinking can be explicitly traced through successive editions of Cotton and Wilkinson, which was long considered the paradigmatic inorganic text (for reference purposes at least, if not necessarily for classroom use, although early editions were explicitly intended to be teaching texts and not reference books [27]). The first edition, published in 1962, takes note of the recent "impressive renaissance" of inorganic chemistry, but opines that "there has been no comprehensive textbook…incorporating the more recent theoretical advances in the interpretation of bonding and reactivity in inorganic compounds," a shortcoming they claim to remedy [28]. Even so, fully 730 of its 950 pages consist of material organized by periodic group, although it is true that their treatment of that material goes beyond simple descriptive chemistry to a much greater extent than preceding texts. This philosophical approach— favoring interpretation over description—was maintained up through the fourth edition [29], which overall was expanded considerably (nearly 1400 pages) over the first even though descriptive chemistry was kept to about the same length (760 pages).

By the fifth edition, however, things changed. There the authors acknowledge "the absence of much theoretical material previously included," for which they offer three reasons. First, the book was getting too long, so *something* had to go; second, they felt that the theoretical background was well covered in texts aimed at a lower level. But third, and most significantly for this discussion, they acknowledge having "become less persuaded of the value of certain types of theorizing. Theories come and go" [30]; so much so that in the sixth edition [31] (completed after Wilkinson's death), descriptive material accounts for 80 % of the total length!

Perhaps the most extreme example of such a contrarian position (calling it "reactionary" might be a little too strong, but not by much) may be found in a

1986 polemic [32] by Jerry Zuckerman, then professor of inorganic chemistry at the University of Oklahoma. Complaining that "Inorganic chemistry is facing an identity crisis…. Inorganic chemists have been coaxed away from their formerly strong, central position based on a monopoly of information on syntheses, reactions and properties of the elements and their compounds by the more ephemeral allure and false sophistication of spectroscopy and theory," Zuckerman called for "a rebirth of descriptive chemistry and its broadest possible exposure to the students." From the evidence collected in this book, I find it hard to accept his assertion that the position of inorganic chemistry had waned in any important sense, let alone his prescription for ameliorating the downward trend he claims to perceive.[4]

So what is inorganic chemistry today, and what is its future? It would be rash to extrapolate too far, but at present there are no visible indications that the field will fragment, or indeed that any wholesale reorganization of chemistry is likely. Proposals for the latter, such as redistributing aspects of organic, physical and inorganic chemistry in favor of new courses and research programs called by names such as synthesis, structure, dynamics, etc., have been put forth with some regularity. Generally, though, these have not taken hold (except perhaps, to some extent, with respect to undergraduate lab courses). As far back as 1969, George Hammond (professor of physical chemistry at Caltech) commented [33]: "For several years, I have been saying that it is a mistake to describe chemistry using a conceptual organization that was created in the last century….even the traditional meanings of the names known to the ingroup of educated chemists are not really good descriptions of coherent fields of reactivity…. My colleague, Professor Harry Gray, is usually called an inorganic chemist, but he states that, 'Inorganic chemistry is a ridiculous field.'"

Ridiculous or not, the field *has* remained coherent—at least in the sense of a large body of chemists choosing to center their activities around organizations and institutions that identify themselves with inorganic chemistry. Bensaude-Vincent and Stengers have observed: "Of all the sciences, chemistry exhibits, it seems to us, a peculiarity in the definition of its territory….How can we assign an identity to a discipline that seems to be everywhere and nowhere at once?" [34]. We would surely find ourselves in comparable difficulty if challenged to come up with a clear and concise definition of inorganic chemistry. But we can get around that problem by understanding that the unity of the field inheres in the association of its members, not in the nature of the research topics that fall under any such heading. The definition offered in Huheey's textbook of some 40 years ago, although at first seeming trivially redundant, strikes me as by far the best and most consistent with

---

[4] It is not impossible that Zuckerman's tongue was partially in his cheek while writing this piece: he closes by quoting, and seemingly endorsing, Thomas Gradgrind's (in)famous exhortation from Dickens's *Hard Times*: "Now what I want is Facts. Teach these boys and girls nothing but facts. Facts alone are wanted in life. Plant nothing else, and root out everything else." Surely Zuckerman understood what Dickens was about in this passage (recall, for instance, that the schoolmaster charged with implementing that program is named McChoakumchild)!

the story told here: "Inorganic chemistry is any phase of chemistry of interest to an inorganic chemist" [35]. The developments that began in the 1950s gave inorganic chemists a sense of intellectual unity that they were happy to associate with; that unity has endured and sustained a field whose identity is defined, not so much by its practices, but rather by its practitioners.

References

1. Woodward WR, Cohen RS (eds) (1991) World views and scientific discipline formation. Kluwer, Dordrecht
2. Nye MJ, Richards JL, Stuewer RH (eds) The invention of physical science: intersections of mathematics, theology and natural philosophy since the seventeenth century. Kluwer, Dordrecht
3. Lemaine G, Macleod R, Mulkay M, Weingart P (eds) (1976) Perspectives on the emergence of scientific disciplines. Mouton & Co, The Hague
4. Lenoir T (1997) Instituting science: the cultural production of scientific disciplines. Stanford University Press, Stanford, CA
5. Dolby RGA (1976) The case of physical chemistry. In: Lemaine G, Macleod R, Mulkay M, Weingart P (eds) Perspectives on the emergence of scientific disciplines. Mouton & Co., The Hague
6. Servos JW (1990) Physical chemistry from Ostwald to Pauling: the making of a science in America. Princeton University Press, Princeton NJ
7. Barkan DK (1992) A usable past: creating disciplinary space for physical chemistry. In: Nye MJ, Richards JL, Stuewer RH (eds) The invention of physical science: intersections of mathematics, theology and natural philosophy since the seventeenth century. Kluwer, Dordrecht
8. Nye MJ (1993) From chemical philosophy to theoretical chemistry: dynamics of matter and dynamics of disciplines, 1800–1950. University of California Press, Berkeley, CA
9. Nye MJ (1996) Before big science: the pursuit of modern chemistry and physics, 1800–1940. Twayne Publishers, New York
10. Barkan DK (1999) Walther Nernst and the transition to modern physical science. Cambrudge University Press, Cambridge, UK, chapter 1
11. Nye MJ (1993) From chemical philosophy to theoretical chemistry: dynamics of matter and dynamics of disciplines, 1800–1950. University of California Press, Berkeley, CA, p 28
12. Nye MJ (1996) Before big science: the pursuit of modern chemistry and physics, 1800–1940. Twayne Publishers, New York, p 20
13. Frickel S, Gross N (2005) A general theory of scientific/intellectual movements. Amer Sociological Rev 70:204–232
14. Collman JP (2012) Personal communication
15. Holm RH (2011) Voices of inorganic chemistry (interview with Richard Eisenberg) on website http://pubs.acs.org/page/inocaj/multimedia/voices.html
16. Valentine JS (2011) Voices of inorganic chemistry (interview with Richard Eisenberg) on website http://pubs.acs.org/page/inocaj/multimedia/voices.html
17. Gray HB (2011) Voices of inorganic chemistry (interview with Richard Eisenberg) on website http://pubs.acs.org/page/inocaj/multimedia/voices.html
18. Lippard SJ (2011) Voices of inorganic chemistry (interview with Richard Eisenberg) on website http://pubs.acs.org/page/inocaj/multimedia/voices.html
19. Gray HB (2013), personal communication

20. Law J (1976) Theories and methods in the sociology of science: an interpretive approach. In: Lemaine G, Macleod R, Mulkay M, Weingart P (eds) Perspectives on the emergence of scientific disciplines. Mouton & Co., The Hague
21. Morris PJT (2001) Between the living state and the solid state: chemistry in a changing world. In: Reinhardt C (ed) Chemical sciences in the 20th century: bridging boundaries. Wiley-VCH, Weinheim, p 199
22. Norris JF (1921) A textbook of inorganic chemistry for colleges. McGraw-Hill, New York
23. Bailar JC Jr. (1989) A history of the Division of Inorganic Chemistry, American Chemical Society. J Chem Ed 66:537–545
24. Moeller T (1952) Inorganic chemistry: an advanced textbook. John Wiley & Sons, New York
25. Shriver DF, Atkins P, Langford CH (1994) Inorganic chemistry, 2nd ed. W. H. Freeman, New York
26. Wood CW, Holliday AK (1960) Inorganic chemistry: an intermediate text. Butterworths, London
27. Cotton FA, Wilkinson G (1966) Advanced inorganic chemistry: a comprehensive text, 2nd ed. Interscience Publishers, New York, p ix
28. Cotton FA, Wilkinson G (1962) Advanced inorganic chemistry: a comprehensive text. John Wiley & Sons, New York, p v
29. Cotton FA, Wilkinson G (1980) Advanced inorganic chemistry: a comprehensive text, 4th ed. John Wiley & Sons, New York
30. Cotton FA, Wilkinson G (1988) Advanced inorganic chemistry: a comprehensive text, 5th ed. John Wiley & Sons, New York, p v
31. Cotton FA, Murillo CA, Bochmann M, Grimes RN (1999) Advanced inorganic chemistry: a comprehensive text, 6th ed. John Wiley & Sons, New York
32. Zuckermann JJ (1986) The coming renaissance of descriptive chemistry. J Chem Ed 63:829–833
33. Russell CA (1976) The structure of chemistry. Open University Press, Milton Keynes, p 10
34. Bensaude-Vincent B, Stengers I (1996) A history of chemistry (translated by D. van Dam). Harvard University Press, Cambridge, MA, p 4
35. Huheey JE (1972) Inorganic chemistry: principles of structure and reactivity, 1st ed. Harper & Row, New York, p 1

Notes on Quantitative Methodology

Lies, damned lies, and statistics
Attributed [probably incorrectly (http://www.york.ac.uk/depts/
maths/histstat/lies.htm)]

by Mark Twain to Benjamin Disraeli

1. Faculty numbers. The two years chosen for comparison, 1953 and 2003, correspond to the first publication of the DGR and the last to be issued in hard copy (using search routines on the later online versions might be thought to be easier than manually going through a book, but spot checks showed they are not reliable). For consistency the survey includes only Ph.D.-granting chemistry departments in the US (later editions include large numbers of Masters-only programs as well as Canadian Ph.D. programs) and excludes (where clearly indicated) emeriti and non-tenure track faculty. Those who identified themselves with a recognized subfield of inorganic chemistry (organometallic, bioinorganic, etc.) were included in the count of inorganic chemists, as were any who gave no identification at all (not many of these) but whose research interests and publications were closely related to others in the field. There may be a small amount of overcounting resulting from ambiguous cases, but even if so, it should apply equally to the "before" and "after" numbers.

2. National Academy of Science membership. The NAS webpage[1] provides a complete list of members—active, emeriti, and deceased—which is searchable by interest section; each individual listing gives the year of election and (usually) a statement or biographical piece that identifies the member's specialization within chemistry. Unfortunately, a significant number of entries, especially the older ones, do not include any interest section, and hence are not found by searching on the latter. Indeed, a search on chemistry produces no inorganic chemist elected before 1960; but the one we met in Chap. 4, Don Yost, was elected in 1944, as can be determined by searching under his name instead (although his brief biographical blurb describes him as a physical chemist!). Thus the data shown in Fig. 3.5 is incomplete, but hopefully the omissions will not have generated any systematic errors—there is no obvious reason why chemists from any one subfield

---

[1] http://www.nasonline.org/member-directory/

J. A. Labinger, *Up from Generality*, SpringerBriefs in History of Chemistry, DOI: 10.1007/978-3-642-40120-6, © The Author(s) 2013

should be more likely to be listed with their interest section unspecified than another—so the trends should be reliable.

3. Papers in *JACS*. The titles of all papers in every fourth issue for the years chosen (1953, 1963, 1973, 1983, 1993) were read on the online Table of Contents maintained at the ACS Publications website[2] and assigned to the appropriate subfield. Assignments were frequently ambiguous: it is difficult to choose between organic or physical for a particular paper on kinetics of an organic reaction; between inorganic or physical for one on spectroscopy of an inorganic compound; etc. I used criteria such as the main focus of the work and the subfield with which the authors identified themselves, and did my best to apply them consistently over the years; hence even though the absolute numbers may be somewhat distorted, the trends should not. There may also be some concern over whether the sample was sufficiently large. Especially in earlier years, there was a tendency to publish a number of papers from one author in a single issue. (For example, issue #1 of 1953 contains significantly more inorganic papers than any other issue I scanned from that year, but over two-thirds of them were from a single source, studies on boron hydrides by Schlesinger.)

As a check on both these potential problems, we can compare some of my numbers to those based on *JACS*' own classifications. Until around 1970, papers in the *JACS* Table of Contents were grouped and identified by subfield. These designations are not readily recoverable: those headings are *not* shown in the online Table of Contents, and most libraries that still have sets of hard copies (including our holdings at Caltech) apparently had the Table of Contents pages removed before binding. However, I was able to locate a set that did have some (not all, unfortunately) of the Tables of Contents at Indiana University, and a student there, Margaret Janz, was kind enough to assemble the statistics. During the period 1952–1959 the fractions of published full papers (communications were not classified in the Table of Contents) classed as organic, physical and inorganic averaged to 60, 25, and 10 % respectively (biochemistry made up the small balance). Those numbers remain fairly constant across that period—the standard deviation for the inorganic fraction is about 0.7 %—and there is no discernable trend at all. In contrast, for the years 1963–1969 inorganic shows a substantial increase—the corresponding numbers are 54, 25 and 16 %—as well as a strong upward trend, from closer to 10 % at the beginning of the period to over 20 % at the end. Comparing to the numbers in Fig. 3.6, we see that the sampling of data for 1953 (63, 18, and 12 %) agrees quite well with the larger data set for the 1950s, while that for 1963 (51, 16, and 25 %) appears to overrepresent inorganic chemistry somewhat. That may be in part a consequence of the small sample size, but a more important factor is that, in the survey I carried out, the fraction of inorganic communications was significantly higher than that of inorganic full papers. Since novelty and urgency are criteria for acceptance as a communication

---

[2] http://pubs.acs.org/journal/jacsat

in *JACS*, that discrepancy may well be another indicator of the up-and-coming status of inorganic chemistry in the early 1960s. A more extensive detailed analysis comparing frequency of communications versus full papers for different subfields might well be another valid metric, but I have not yet carried out such an exercise. In any case, it seems clear that the procedure used to construct Fig. 3.6 *does* capture reality well enough to establish firmly the change over the decades, and to locate the beginnings of that change sometime close to 1960.

4. Presentations at national ACS meetings. Full programs for all meetings (there are only a couple of lacunae) may be found in the online archives of *Chemical and Engineering News*,[3] and counting the number of listings under each Division is mostly straightforward. The only complication arises for pre-1957 meetings, when there was no separate Inorganic but rather a joint Physical and Inorganic Division. In those programs many of the individual sessions were specifically characterized as inorganic or physical, and hence unproblematic; for the others, individual papers were assigned to one or the other using the same criteria as for papers in *JACS*. In addition, frequently meetings feature symposia jointly sponsored by two or more Divisions. In such cases all the papers were assigned to the Division under which they were listed, rather than trying to apportion them. Any distortions thus introduced will surely cancel out over time.

5. Academic lineages. Faculty listings in the ACS Directory of Graduate Research include institution and year of Ph.D. degree and, often, postdoctoral fellowships (the latter are not always complete), from which it is *usually* possible to identify (by locating joint publications, in particular) the graduate or postdoctoral mentor, and thus trace backwards to determine whether or not academic descent starts with one of the six founding groups. Obviously that process is more difficult as the chain becomes longer. The year 1983 was chosen for the exercise, as a point early enough so that factor was not *too* painful, but late enough that the large majority of active faculty *could* have gotten their training under a founder or descendent of those groups. As noted in Chap. 5, 285 of the 566 inorganic faculty listed did so. There were 16 more that I could not trace by this procedure: they *may* belong in those lines of descent, but I found no revealing publications. The number is sufficiently small that the extensive effort of direct inquiry that would be needed to confirm them did not seem worthwhile.

---

[3] http://pubs.acs.org/journal/cenear

About the Author

Jay Labinger is a California native, born in Los Angeles in 1947. He was an undergraduate at Harvey Mudd College, and received his Ph.D. (in inorganic chemistry, of course) at Harvard University in 1974. After a postdoctoral stint at Princeton University, he held successive positions in academia (University of Notre Dame) and industry (Occidental Petroleum, ARCO) before coming to Caltech in 1986, where he is Administrator of the Beckman Institute and Faculty Associate in Chemistry. His chemistry research has been focused in the areas of organotransition metal chemistry and energy-related catalysis. Many of his contributions have taken the form of mechanistic explanation of transformations that are potentially valuable in the energy sphere; these include oxidative coupling of methane, selective oxidation of alkanes by soluble metal complexes, and conversion of methanol to a high-octane hydrocarbon. He was elected Fellow of the American Association for the Advancement of Science in 2009.

For the last twenty years or so, he has also been active in scholarship on the borders between science and the humanities, writing on topics such as science and literature, controversial episodes in the history of chemistry, and the "Science Wars." He co-edited (with Harry Collins) the book *The One Culture* (2001), a conversation-in-print between scientists and scholars of science. He is a past president of the Society for Literature, Science and the Arts.

J. A. Labinger, *Up from Generality*, SpringerBriefs in History of Chemistry, DOI: 10.1007/978-3-642-40120-6, © The Author(s) 2013